T0348733

Thin Film Micro-Optics

New Frontiers of Spatio-Temporal Beam Shaping

Thin Film Micro-Optics
New Frontiers of Spatio-Temporal Beam Shaping

Ruediger Grunwald

Max Born Institut for Nonlinear Optics and Short Pulse Spectroscopy
Berlin, Germany

Amsterdam • Boston • Heidelberg • London • New York • Oxford
Paris • San Diego • San Francisco • Singapore • Sydney • Tokyo

ELSEVIER

Elsevier
Radarweg 29, PO Box 211, 1000 AE Amsterdam, The Netherlands
The Boulevard, Langford Lane, Kidlington, Oxford OX5 1GB, UK

First edition 2007

Library of Congress Cataloging-in-Publication Data
A catalog record for this book is available from the Library of Congress

British Library Cataloguing in Publication Data
A catalogue record for this book is available from the British Library

ISBN-13:   978-0-444-51746-3
ISBN-10:   0-444-51746-4

Cover illustration:
Ultrashort wavepacket in space and time detected by a multichannel autocerrelator
with thin-film beam shaper (see Fig. 184).

For information on all Elsevier publications
visit our website at books.elsevier.com

Transferred to digital print 2007
Printed and bound by CPI Antony Rowe, Eastbourne

07 08 09 10 11   10 9 8 7 6 5 4 3 2 1

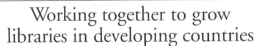

FOR SABINE,
       LINN,
       CHRISTOPH,
       AND MY MOTHER.

PREFACE

Thin-film micro-optics stands for a concept that unifies the particular properties of miniaturized optical systems with the specific advantages of thin optical films. Structured dielectric, metallic or compound layers can be fabricated by simple and flexible procedures. Significantly reduced aberrations at very small angles meet new spectral degrees of freedom, low dispersion and minimal absorption. Thus, improved functionality can be realized which opens interesting novel prospects for the transformation of complex light fields in space and time. High-power, ultrashort-pulse and short-wavelength applications are enabled. In this sense, thin-film micro-optics represents a new and fruitful branch at the family tree of optics.

The author had the luck to be involved in a number of projects in micro-optics and laser beam shaping during a decade full of pioneering activities. Remember the stimulating atmosphere of the EOS Topical Meetings at NPL in Teddington under leadership of Mike Hutley, the Micro-optics Workshop in Erice, Sicily, in November 1996, or MOC and DOMO conferences, to mention only a few of the events. In the maelstrom of booming microelectronics and telecommunication industries, the necessary technological basis for a reliable mass production of microlens arrays, gradient index (GRIN) lenses and diffractive optical elements (DOE) became accessible. Important stations of this success story like photolithography or resist reflow techniques and characteristic properties of sub-millimeter size optical elements are already described in excellent books written by competent colleagues like Hans-Peter Herzig, Stefan Sinzinger and Jürgen Jahns, Dan Daly, Nicholas Borelli, Maria and Stefan Kufner and others. Beside the main stream of these "conventional" micro-optical technologies, but very close to forefront developments in quantum electronics and photonics, alternative solutions came up recently. It was found that a great variety of layer profiles can be shaped in one- or multistep-procedures by combining conventional, cost-effective vapor-deposition techniques with shading masks placed in the proximity to substrates. With spatially variable multilayers, even spectral amplitude and phase of a light wave can be shaped. For example, micro-gradient AR coatings can be adapted to an incident beam of given spatial and angular intensity distribution. With arrays of thin-film microaxicons, the beam of a femtosecond laser can be split into multiple nondiffractive beams thus allowing for a spatially resolved analysis of pulse duration and wavefront curvature. Recently, such components were successfully used to generate optical X-waves (i.e. propagation-invariant X-shaped spatio-temporal wavepackets) for the first time which are of interest for future robust and secure high-speed information processing systems, nonlinear optics and materials processing.

The aim of the book is to give a comprehensive overview of the specific properties of thin-film micro-optics, specific degrees of freedom for thin-film design, problems of the numerical simulation of beam propagation, different fabrication and characterization techniques, and selected fields of applications. The examples are mainly concentrated on continuous-relief structures in refractive, reflective and hybrid design for shaping and characterizing high-power ultrashort-pulse laser beams.

Next generation micro-optics, will be closely connected to all the remarkable trends in optics, laser physics, optoelectronics and information technology. The integration of adaptive components, the combination with nanooptical devices, photonic crystals or metamaterials, the exploitation of nonlinear properties, the implementation of biologically inspired architectures or the advance to ultra-precision components have to stay subjects of continuing work. May this book be a useful tool for colleagues and students and a modest contribution on the sometimes stony, but ever fascinating road to future optical technologies.

Rüdiger Grunwald

Berlin, 2006

ACKNOWLEDGEMENTS

First of all, the author has to thank Uwe Neumann (Berlin) for critically reading and layout of the manuscript and Monika Tischer (MBI Berlin) for her technical assistance.

The main part of the activities behind the book was performed at the Max Born Institute for Nonlinear Optics and Short-Pulse Spectroscopy (MBI) in Berlin. Therefore, I like to thank all the colleagues from MBI, in particular director Prof. Thomas Elsaesser, for supporting the work. Important input I further owe joint projects with the Bremen Institute of Applied Beam Technology (BIAS) where advanced measuring techniques were developed under the supervision of Prof. Werner Jüptner.

Some of the most recent results were enabled by the expertise of Dr. Günter Steinmeyer and Gero Stibenz in compressing and analyzing ultrashort pulses. Special thanks go to my colleague Uwe Griebner (Berlin) who successfully survived tons of paper, thousands of telephone calls and discussions of crazy ideas. It was a pleasure to collaborate with Volker Kebbel (Bremen) who propagated X-waves through his computer and spent days and nights in common experiments and brain storming. The same can be said about Siegfried Woggon (Karlsruhe) who did a very qualified job in a rather difficult time. The art of shadow-mask vacuum deposition was cultivated in the team of Dieter Schäfer (Berlin), in particular by Hans-Joachim Kühn and Rudi Ehlert. By directing my attention to edge effects of multilayers, Günther Szczepanski initiated the application of three-dimensional masks to the deposition of thickness modulated layer components. Johannes Heinrich Hertz (Bohnsdorf) gave me practical lectures about beauty and style in experimental physics. Horst Schönnagel (Berlin) told me about Bessel beams for the first time. I am grateful for his patient answers on stupid questions about everything in optics. With Ralf Koch (Stockholm) and Michael Kreitel (Berlin) we had enlightened projects. I like to mention beneficial working stays in Canada in the groups of Michel Piché (Quebec) and Peter R. Herman (Toronto). Michel was significantly involved in developing concepts on nondiffracting beams. Thanks to him also for the great hospitality and the maple sirup of his old father.

For stimulating discussions, technical support and/or any other kind of helpful collaboration I thank A. Bärwolff, I. M. Besieris (Blacksburg), M. Bock, K.-H. Brenner (Mannheim), E. Büttner, C. Conti (Rome), P. Corkum (Ottawa), S. Doric (Quebec), A. Duparré (Jena), G. Erbert, M. Ferstl, A. Fischer, D. Fischer, M. Fortin, M. Friedrich, J. Gähde, P. Glas, W. Goleschny, R. Güther, P. Hamm (Zürich), H.-J. Hartmann (Hannover), H.-H. Heinrich, J. Herrmann, H. E. Hernández-Figueroa (Campinas), H.-P. Herzig (Neuchâtel), T. Hessler (Sarnen), J. Jahns (Hagen), F. Kneubühl† (Zürich), W. Karthe (Jena), E.-B. Kley (Jena), H. Knuppertz (Hagen), K. Kolasinski (London), F. Krausz (Garching), K. Krüger, A. Kummrow (Berlin), S. Langer, J. Leger (Minneapolis), M. Lehmann, U. Leinhos (Göttingen), J. Li (Toronto), A. Lohmann (Erlangen), E. Lüder (Stuttgart), C. Lukas, K. Mann (Göttingen), H. Mischke, E. McGlynn (Dublin), R. Menzel (Potsdam), J.-P. Mosnier (Dublin), J.-P. Müller, R. Müller, M. R. Murphy, M. Naumann, J.-L. Néron, S. Nerreter, E. T. J. Nibbering, E. Nirschl (Regensburg), W. Osten (Stuttgart), B. Ozygus, H.-J. Paetzold, E. Pawlowski (Mainz), R. Piestun (Boulder), J. Popien, T. Poßner (Jena), E. Recami (Bergamo), W. Rehak, K. Reimann, W. Reinecke, C. Reschke, A. Richter (Wildau), M. Rini (Berkeley), D. Ristau (Hannover), R. Rondt, A. Rosenfeld, G. Rousseau (Quebec), S. Ruschin (Tel Aviv), P. Saari (Tartu), T. Sandrock (Bonn), R. Sauerbrey (Dresden), B. Schäfer (Göttingen), A. Schindler (Leipzig), M. Schmidbauer, G. Seewald, J. Sheridan (Dublin), J. Schwider (Erlangen), W. Seeber (Jena), G. Seewald, A. M. Shaarawi (Cairo), W. Singer (Oberkochen), K. Schwenkenbecher, S. Schwirzke-Schaaf, H. Tiziani (Stuttgart), J.-W. Tomm, R. Trebino (Atlanta), F. Tschirschwitz, A. Tünnermann (Jena), H. Weber, G. Wernicke, B. Wilhelmi (Jena), H.-H. Witzmann, M. Wörner, D. Wulff-Molder (Altenholz), L. Xiang (Hangzhou), H. Zappe (Freiburg), M. Zamboni-Rached (Campinas), H. Zimmermann, and all who are not mentioned explicitely.

Research projects and international collaboration were financially supported by BMBF (16SV275/2, 01M3025C, 13N7474/7, CAN 00/016, CAN 02/15) and DFG (Gr 1782/2-1, Gr 1782/7-1).

Thin-film microoptics - New frontiers of spatio-temporal beam shaping

TABLE OF CONTENTS

LIST OF ABBREVIATIONS

2D	two-dimensional
3D	three-dimensional
AFM	atomic force microscopy
AR	anti-reflection
ASHS	axicon-based Shack-Hartmann sensor
BASIC	backside-coated mirror
BBO	Beta-Barium Borate (crystal)
BGB	Bessel-Gauss beam
BK7	crown glass: "Bor-Kron-Glastyp Nr. 7" (Schott, Germany)
BPM	beam propagation method
CD	compact disk
CEP	carrier-envelope phase
CM	chirped mirror
CPA	chirped pulse amplification
cw	continuous wave
DCM	doubly chirped mirror
DLP^{TM}	Dynamic Light Processing^{TM}
DMD^{TM}	Digital Micromirror Devices^{TM}
DOE	diffractive-optical element
DOF	depth of focus
DVD	digital versatile disk
EE	encircled energy
ESCA	electron spectroscopy for chemical analysis
FAC	fast-axis-collimation
FIR	far infrared
FOV	field of view
FROG	frequency resolved optical gating
FWHM	full width at half maximum
GD	group delay
GDD	group delay dispersion
GE	General Electrics (company)
GRIN	gradient index
GRM	graded reflectance mirror
GRMMA	graded reflectance micro-mirror array
GTI	Gires-Tournois interferometer
HASHS	hyperspectral axicon-based Shack-Hartmann sensor
HDTV	High Definition Television
HR	high reflectivity
IR	infrared
ISO	International Organization for Standardization
LBO	Lithium Borate crystal for SHG
LC-SLM	liquid-crystal-based spatial light modulator
LCVD	laser chemical vapor deposition

LED	light emitting diode
LIGA	lithography with galvano-plating and molding
LPVD	laser physical vapour deposition
LVDS	low voltage differential signaling
MAMA	micro-axicon matrix processor
MEA	micro-etalon array
MEMS	micro-electro-mechanical system
MFL	micro-Fresnel lens
MLA	micro lens array
MOEMS	micro opto electro mechanical system
MOPA	master oscillator and power amplifier system
MOVPE	metal-organic vapor phase epitaxy
MTF	modulation transfer function
NA	numerical aperture
NEMS	nano-electro-mechanical system
NOEMS	nanooptical-electro-mechanical system
OPD	optical path difference
PC	polycarbonate
PES	polyether sulphone
PI	polyimide
PMMA	polymethyl methacrylate
PSF	point spread function
PVD	physical vapor deposition
R	Röhm (company)
RIE	reactive ion etching
rms	root mean square
SAC	slow-axis collimation
SBP	space-bandwidth product
SEM	scanning electron microscopes
SHG	second harmonics generation
SHS	Shack-Hartmann sensor
SI	Système International d'Unités, international unit system
SLM	spatial light modulator
SNOM	scanning near-field optical microscopy
SR	Strehl ratio
TFI	tilted-front-interface chirped mirror
TFM	thin-film microlens
UV	ultraviolet
VCSEL	vertical-cavity surface-emitting laser
VRM	variable reflectance mirror
VUV	vacuum-ultraviolet
WPM	wave propagation method
XGA	Extended Graphics Array
YAG	yttrium aluminum garnet

Chapter 1

INTRODUCTION

"I call our world Flatland..."

Edwin Abbott Abbott
[1884_1]

The most recent advances in optics are characterized by ever increasing resolution in time, space and spectrum which are enabled by ultrashort-pulse lasers, micro- and nanooptical systems or high-resolution spectroscopy. Thus, laser materials processing of unprecedented precision can be performed, ultrafast and massive-parallel information transfer can be realized and the interaction of photons with physical, chemical or biological matter can be well localized in space and time.

In particular, the miniaturization of optical components was a first important milestone on this road. However, specific physical limitations given by spectral dispersion, diffraction or transmission proved to be drawbacks for light confinement and its applications to shaping, propagating and characterizing wavepackets at ultrashort pulse duration, ultrabroad spectral bandwidth or extremely high photon energies. To overcome these limits, new and unconventional approaches for the design of microoptical components as well as system architecture were urgently needed. As it will be shown in this book, essential progress became possible by *combining* the state of the art of *miniaturized optical elements* with specific advantages of another well-established branch, the *thin-film technology*. After introducing the basic properties of classical microoptics and thin-film optical devices, the new and promising concept of *thin-film microoptics* will be presented.

The problems of design, fabrication, characterization and operational principles will be discussed in detail. The advantages of the thin-film microoptical approach will be demonstrated by very recent results of ultrashort-pulse and vacuum ultraviolet beam shaping experiments. Particular emphasis will be laid on the generation of well-localized, pseudo-nondiffracting few-cycle wavepackets which may give an idea of the great potential of thin-film microoptics to open new frontiers of spatio-temporal beam shaping. With the generation of needle-shaped beams ("*needle beams*") of extremely small conical angles, a new type of limited-diffraction wave phenomena is available which enables to create extraordinarily extended focal zones. Needle beams can be regarded to be a real physical approximation of rays bridging the gap between geometrical and wave optics in a certain way.

By briefly striking further tantalizing thin-film microoptical approaches like multilayer microoptics, thin-film nanostructures or low-dispersion design, an outlook is dared to the future of thin-film microoptics.

Chapter 2

MICRO-OPTICS

2.1. The concept of microoptics

What in fact is microoptics? Some years ago, the answer was relatively clear. In a rather simple picture, microoptics was defined as (a) miniaturized optical components with typical dimensions on sub-millimeter scale and (b) the specific performance of optical systems comprising such elements [1984_1, 2003_3]. By the major principle of operation, one can differ between refractive, reflective, diffractive or hybrid microoptical components (see [1994_1, 1997_1, 1997_2, 1999_1, 2000_1, 2001_4]). Because of the recent technical progress, however, the face of microoptics was rapidly changed. Nanooptics, microfluidics, micro-electro-mechanical systems (MEMS), microstructured fibers, liquid crystals or digital mirror devices (to mention only a few) are integrated in complex optical systems together with various microoptical building blocks. Miniaturized components (lenses, gratings, mirrors, prisms and waveguides) as key components of highly complex, intelligent systems are of increasing interest for the spatio-temporal control of light fields (image, laser beam) [1994_1, 1995_18, 1996_21].

The central role that microoptics plays within the manifold of highly dynamic, overlapping fields of micro- and nanosystem technologies is adumbrated in the simplified scheme in Fig. 1. Because of the large diversity of design approaches, it is urgently necessary to concentrate here on selected new topics of particular relevance for spatio-temporal beam shaping while other likewise interesting topics will only be touched very briefly.

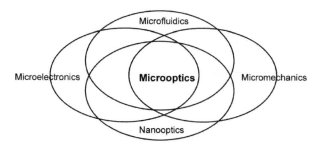

Fig. 1. The central role of microoptics as an enabling technology in the overlapping fields of micro- and nanosystem technologies.

Microoptics is all but a new concept. About 500 million years ago, compound eyes of crustaceans living in the oceans were successfully designed by the powerful selection rules of evolution, even several times [1975_1, 1988_8, 1996_22, 1997_29, 2002_2]. Beautiful color effects at dew droplets or butterfly wings and fish scales were well-known long before people started to systematically develop microoptical technologies. The pocket microscopes of the Huygens brothers used glass beads as miniaturized ball-lenses and may be regarded as early microoptical instruments. The *"big-bang" of microoptics*, however, surely began some hundred years later with the mass-production of sophisticated semiconductor-based electronic devices and low-loss optical fibers for photon-based telecommunication and networking.

High-quality microlenses were needed to collimate and to couple light from sources into optical fibers or waveguides and to focus it onto detectors [1991_1]. As a first milestone in microoptics history, gradient index (GRIN) lenses were already developed in the fifties of last century [1984_1]. Ion exchange and diffusion procedures had to be established. During the following decades, the full potential of available technologies was applied to the fabrication of optical microstructures. By adapting lithographic and etching technologies from semiconductor industry, high-precision optical treatment, direct writing with focused laser or ion beams, transferring structures into polymers and glasses and developing replication methods, billion-dollar markets like displays, CD- and DVD-players or digital cameras could be served (for a comparison of different methods see, e.g., 1997_1, 1999_1). Large-area arrays of refractive, diffractive or hybrid microlenses can be produced with fairly high reproducibility. The progress in theoretical optics combined with the ever-increasing capacity and speed of computer hard- and software on one side and the enormous accuracy of microstructuring techniques enables now for the design of even sophisticated diffractive-optical elements (DOE) of very high complexity. *Subtractive, modifying or additive procedures and combinations* of them can be used as basic technologies for the fabrication of microoptical components [1997_5]. Compared to "conventional" macro optics, micro-optical systems have specific advantages and disadvantages with respect to a spatio-temporal beam shaping. Scale-variant properties have carefully to be taken into account for the design of components and system architectures. Extended system functionality like parallel processing or coherent coupling of multiple channels can be realized by *arrays* of miniaturized optical elements [1991_3].

In principal, microoptical systems enable all optical transformations of light fields that are also known from macroscopic-size optical systems. The main application fields are schematically drawn in Fig. 2 [2005_5].

Imaging microoptical systems are mainly used for image relay, magnification, collimation, focusing and filtering. Typical examples are the collimation of laser diodes or LED [1989_10, 1989_11, 1991_3, 1991_11, 1993_10, 1994_10, 1994_30, 1994_32,

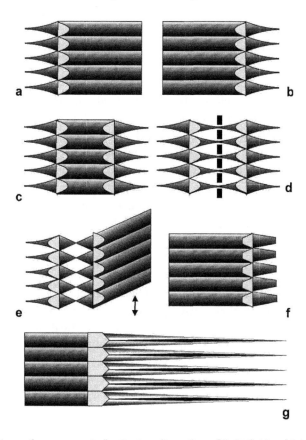

Fig. 2. Frequently used arrangements for the transformation of light fields with imaging and non-imaging microoptical array systems: (a) collimation, (b) focusing, (c) imaging / coupling, (d) 4f-imaging with filter array, (e) beam deflection, (f) light collection, (g) axial beam shaping. The shown configurations are represented by (but not restricted to) refractive-type components (after [2005_5]).

pp. 199-208, 1995_1, 1995_15], focusing (drilling of holes, coupling into fiber bundles) [1994_11, 1986_2, P4] or beam steering for projection [1989_15, 1994_8, 1997_31]. Other applications are coherent coupling, redistribution, homogenization and shaping of light distributions (Gaussian, flat-top, circular, annular etc. [1989_16, 1995_20, 1995_21, 1995_22]) for laser illumination (lithography), materials processing [1994_7] or laser medicine. The imaging properties enable to build up free-space and 3D-integrated waveguiding couplers and multiplexers for communication systems and optical computers [1987_4, 1991_19, 1991_20, 1994_14, 1994_33, 1994_34]. Further important applications are array generation [1991_25, 1992_8], copy protection [1997_32] or focus sensing in cameras by passive triangulation [1995_19]).

Nonimaging systems are applied to beam steering, light collecting systems and axial beam shaping. The microscopic dimensions of the components result in size-dependent effects of different nature. Specific scaling laws are mainly related to diffraction and aberrations, but also to simple geometry and information theory (see 2.2.1.). A typical domain for microoptical applications are any kind of multichannel architectures because of the high possible density of channels in space and the capability of parallel or space-variant processing [1992_9]. Arrays of microoptical elements like zone-plates [1993_11] enable to extend image sensing techniques to the third dimension. Spectral and angular selectivity is achieved by multilayer or diffractive structures.

With addressable multichannel devices, additional degrees of freedom for a spatio-temporal beam shaping get accessable. A well-known application is the space-to-time shaping of ultrashort-pulse lasers by controlling the spectral phase and/or amplitude with spatial light modulators (SLM) [1993_2] which can be optimized with evolutionary algorithms [1997_6]. Recently, Ti:sapphire laser pulses in 5-fs range [2001_5] and below [2002_3] were generated with SLM-type temporal beam shapers.

In the frame of laser beam shaping at extreme parameters with respect to the pulse duration, angular distribution, wavelength or spectral bandwidth, "classical" microoptical systems suffer from inherent limitations, e.g. by dispersion, diffraction or particular aberrations. By exploiting the advantageous properties of layer structures in the concept of thin-film microoptics, further improvements of the system performance are possible as it will be shown later on. First of all, however, let us take a closer look at the fundamentals of microoptics.

2.2. Microoptics and macrooptics

2.2.1. Scaling laws for a size reduction

In comparison to conventional macrooptical systems, microoptical elements have typical aperture diameters which are about two orders of magnitude smaller. The size reduction enables for systems of high compactness, integrability and flexibility [1997_1]. On the other hand side, it results in specific advantages and disadvantages with respect to the mechanical and optical system demands. The dependence of the optical performance of microlenses on their size can be described by the following *scaling laws* (fitting monochromatic and linear optical conditions) [1994_19]. For a geometrically similar size reduction of imaging lenses, the parameters *focal length*, *lateral aberrations* and *travel time* (i.e. the time for monoenergetic photons to completely pass through the lens) are scaling down linearly [1998_1]. The *optical resolution* and the *depth of focus* (DOF) remain invariant, whereas the image field size depends on the square of the scale parameter (ref. [1997_2], pp. 13-22). The scaling of the *space-bandwidth product* (SBP) [1997_7, 1998_1] depends on the particular contributions of aberrations and diffraction [1995_14]. For aberration-free microoptics, the SBP scales quadratically with the system dimensions. Because of the cubic scaling of the volume, the *weight* of microlenses also reduces proportional to the third power of the scaling factor. For engineering systems with fast moving parts like high-speed scanners, one-way detectors or space applications, these aspects are of great practical importance. Finally, it should be mentioned that the influence of local fabrication errors (profile, rim, scratches), material inhomogeneities and contamination (dust) is more critical to smaller dimensions so that quality standards similar to that of microelectronics clean rooms are necessary. With shorter wavelengths and smaller feature size, the demands on the precision of substrates, spacers, holders and each kind of micro-scale accessories are increasing as well.

2.2.2. Diffraction and Fresnel number

In general, the propagation of light can not be separated from diffraction effects. For optical systems with effective apertures of macroscopic dimensions (i.e. with diameters large against the wavelength), the laws of geometrical optics and the picture of rays can be applied in good approximation [2006_4]. For the transformation of light fields by components of microscopic size (i.e. with diameters down to only a few wavelengths or less), diffraction effects have carefully to be taken into account [2006_3].

A useful measure for the contribution of the diffraction to the performance of optical systems is the well-known *Fresnel number* [1984_4]

$$F(z) = \frac{d^2}{4\lambda z} = \frac{a^2}{\lambda z} \tag{1}$$

which is (in monochromatic case) a function of the aperture diameter $d = 2a$, the wavelength λ and the axial distance z from the aperture [1998_2, p. 117]. The following two special cases are of particular importance:

$$F(z) \ll 1 \quad (Fraunhofer\ diffraction) \qquad (a)$$

$$F(z) \geq 1 \quad (Fresnel\ diffraction) \qquad (b)$$

In the Fraunhofer regime (a), diffraction plays a major role, whereas it is less important in the Fresnel regime (b). Figure 3 shows the Fresnel number as a function of distance at realistic parameters for low-NA thin-film structures (mention the logarithmic scale of the ordinate axis!). Obviously, this case corresponds the to the Fresnel regime (b).

For a given distance and wavelength, a doubling of the size of the aperture results in quadrupling $F(z)$. The diffraction, vice versa, increases quadratically with reduced

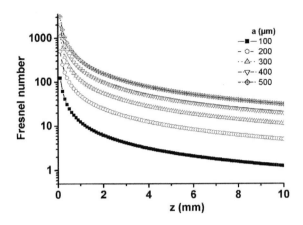

Fig. 3. Fresnel number as a function of the distance z for low-NA structures (a - radius of the aperture, parameter a - radius of the aperture, wavelength fixed at $\lambda = 800$ nm).

aperture size. To determine the Fresnel number $F(f)$ of a focusing microlens, the distance z in equation (1) has to be replaced by the focal distance f.

$$F(f) = \frac{d^2}{4\lambda f} = \frac{a^2}{\lambda f} \qquad (2)$$

The indication of a single Fresnel number is only adequate to rotationally symmetric apertures and monochromatic illumination. For asymmetric elements like cylindrical lenses of rectangular or elliptical cross-section, two or more relevant Fresnel numbers exist. Arrays can be described by effective Fresnel numbers including the diffraction of solitary elements (small scale) as well as the diffraction of the whole array (large scale).

For polychromatic illumination (e.g. in the case of ultrashort pulsed lasers, broadband LED or supercontinuum radiation), the spectral dependence of the diffraction has to be taken into account. It reduces the effective bandwidth by narrowing the spectral transfer functions and causes unwanted channel cross talk and losses (see also 2.9.). In particular, it makes the design of broadband diffractive optical elements (DOE) a challenging task. On the other hand side, diffraction can be used for coupling, filtering and beam shaping.

In the case of DOE, photonic crystals or metamaterials, substructures can be comparable to or even smaller than the light wavelength. For an optimization of such structures and a detailed understanding of light propagation, an advanced theoretical description of diffraction is essentially [1991_13, 2003_16].

2.3. Types of microoptical components

Microoptical components can be sorted in diverse classification schemes (e.g. with respect to shape functions, arrangements, optical functionality, spectral range, material etc.). A selection of some of the most important types can be found in Fig. 4.

Because of the prior importance of the beam shaping *functionality* (and keeping in mind above mentioned relations between aperture size and diffraction), it seems to be meaningful to distinguish between

- refractive,
- reflective,
- diffractive, and
- hybrid

microoptical components (compare [1997_1], pp. 1-29). In all cases, a shaping of a wave field is obtained by modifying the optical path length (phase shift), wavefront tilt (phase gradients) and/or direction (reflection and phase gradients).

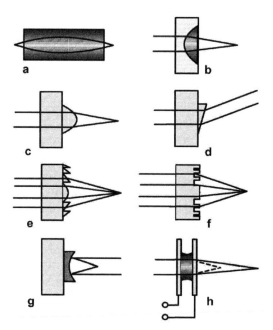

Fig. 4. Basic types of microoptical components: (a) and (b) refractive gradient index (GRIN) lenses (GRIN rod lens, planar GRIN lens), (c) and (d) refractive relief elements (microlens, microprisma) (e) refractive micro-Fresnel-lens, (f) diffractive microlens, (g) reflective components (concave micromirror), (h) variable focus microlens (electrically tunable liquid crystal lens).

Lower chromatic aberrations and higher efficiencies of beam shaping are obtained with refractive components (see also [2006_3]). The degrees of freedom for beam shaping are maximal for DOE. However, applications of refractive components and DOE in ultrashort-pulse domain are seriously limited by spectral dispersion and diffraction. Reflective components require sophisticated off-axis design and often suffer from shading effects and strong asymmetric aberrations. In hybrid components, two or more of the basic qualities are combined with each other either to extend the functionality or to correct for the aberrations.

Refractive structures exploit different approaches. Gradient-index (GRIN) lenses are based on a radial change of the refractive index that results in a radially variable tilt of the wavefront (Fig. 4a,b) [1980_1, 1984_1]. Relief lenses (Fig. 4c) or prisms (Fig. 4d) refract the local wavefront according to Snellius law. Fresnel-lenses and similar parcellated refractive structures (Fig. 4e) consist of substructures of radially variable size (piecewise simulating the phase profile of thicker lenses, compare [1999_1], pp. 6-10). The wave-optical features of Fresnel lenses are relevant as well so that they can be placed in a transition field between refractive and diffractive optics [1995_14] (see paragraph 2.4.4.).

By comparing the particular *mode and quality of transforming light fields*, microoptical elements can further be divided in imaging and nonimaging systems (see Fig. 2). Well-known nonimaging systems are light concentrators (e.g. arrays of pyramid structures enhancing the fill factor of a sensor or detector array). This classification scheme, however, is very rough. In general, no clear dividing line between both types of elements can be defined (imagine, e.g., microlenses with high spherical aberrations or microaxicons). A generalized typology could be related to wave aberrations taking into account the *symmetry properties* of microlenses.

With respect to the *degree of integration*, one can group the microoptical elements in free-standing components, integrated optics and planar-optical stacked systems as combinations of both [1988_3, 1997_1, 1997_2, 1998_14, 1999_1]. From point of view of *fabrication and fine structure*, there are the basic types of analog (continuous relief) and binary (discrete two- or multilevel) microoptical elements and combinations of them. Considering the recent progress in design strategies as well as fabrication technologies (e.g. deep-UV and X-ray lithography of structures in few-nm-range) and post-processing (e.g. laser or ion beam smoothing), the borderline between analog and binary structures vanishes as well.

Addressable and non-stationary microoptical components like digital mirrors and tunable lenses (Fig. 4h) are currently of rapidly increasing interest for adaptive optical tasks in image projection, digital photography, automated microscopy or data storage, to mention only a few. The technical approaches reach from the well-known arrays of tilted micromirrors over electrically tuned liquid crystal lenses to pressure-steered liquid-filled polymer micro-balloons or capillaries. This field will continue to offer new and surprising concepts which might be a technological platform for next generation industrial high-technology equipment as well as consumer products.

Last but not least, we have to speak about *thin-film microoptical structures* which play a star role in this book. On one hand side, all the refractive, reflective as well as diffractive features known from conventional microoptical components can also be realized in thin layer systems by appropriate structuring techniques. On the other hand side, the functionality of thin-film microoptics can be very different from other types of microoptics. For example, if we replace a metallic micromirror (as schematically drawn in Fig. 4g) by a microstructured dielectric multilayer system, additional degrees of freedom for the design of spectral phase and amplitude get accessible. Before we discuss all the specific advantages, problems and applications of layer-based components, we like to start with an overview on the optical properties of known and commonly used "classical" types of microoptical structures.

2.4. Refractive microoptics

2.4.1. Specific properties of refractive microoptical components

As already denoted in section 2.3, refractive operation can be obtained by spatially variable refractive indices like in the case of gradient index (GRIN) lenses [1984_1] (Fig. 4a,b) as well as by surface curvatures or tilts of materials of homogeneous refractive index (Fig. 4c,d). In some cases, both types are mixed (e.g. for GRIN-lenses with curved end-faces or GRIN-fiber lenses). In all cases, curved phase fronts result from spatially variable optical path differences (OPD) of the transmitted waves. The curvature-based structures have again to be divided into free-standing single elements (like ball lenses, fiber lenses or cylinder lenses) and integrated elements (curved parts on a surface of a flat substrate, structured fiber end-faces or integrated lenses-on-chip and prisms-on-chip for semiconductor lasers). The particular lenses of microlens arrays are typically (but not in every case) integrated elements. For planar-microoptical systems, elements with off-axis design like toric lenses are designed [1997_22].

Refractive microoptics can exhibit *continuous* ([1997_1], pp. 87-126; [1997_11]) *or discontinuous surface profiles*. A special case of microlenses with discontinuous global profile (but continuous local partial profiles) are *micro-Fresnel lenses* [1987_1] (MFL, Fig. 4e). Rotational symmetric MFL are divided in a central circular lens (zeroth zone) surrounded by annular zones with spacings in analogy to a Fresnel zone plate (see paragraph 2.4.4.).

Like in the case of macrooptics, the wave aberrations are caused by deviations from standard phase distributions (e.g. spherical aberration) and spectral dispersion (chromatic errors). Compared to large apertures, absolute values of phase errors are much more critical. With decreasing element diameter, the influence of diffraction increases quadratically so that we find a coexistence of refractive and diffractive action in many realistic cases (see also [1995_14]). This has to be taken into account in any design considerations.

2.4.2. Gradient index lenses

Gradient index (GRIN) structures consist of glass or polymeric material with index modifications in the volume (3D) or near surfaces (quasi-2D). Frequently used types of GRIN microoptics are rod-lenses (Fig. 4a) with radial index profiles and planar lenses with local modifications of dopand concentrations by diffusion, ion exchange or implant (Fig. 4b) [1995_9, 1984_3]. Refractive index gradients are classified on the basis of *isoindicial surfaces* where characteristic position variables $p(x,y,z)$ remain constant [1983_1]. To simulate the ray propagation in GRIN lenses, ray equations adapted to the particular index distribution have to be solved ([1984_1] and [1999_3], pp. 59-72). One of the most important distribution functions is the *parabolic index profile*. The solutions of the corresponding ray equations are harmonic functions with sine and/or cosine terms ([1999_1], p. 106). For practical applications like coupling [1991_16], the period or pitch P of these oscillating functions is an important reference value because the ratio of P to the rod length L determines the position and number of focal points (schematically in Fig. 5).

To collimate the beams of point-like light sources like fibers, waveguides or diode lasers, quarter-pitch lenses are used. Line-shaped or astigmatic sources as well as high NA beams have to be collimated or coupled with GRIN structures of cylindrical geometry, adapted systems of GRIN lenses or combinations with other microoptical components (typically two or more). With half-pitch lenses (Fig. 5a,b), a transformation of a point source to a collimated beam or vice versa is obtained. If half-pitch lenses are used for imaging, an inverted image results. A coupling of a point source to a point detector or fiber and an image transfer to an upright image is possible with an integer

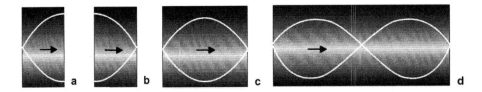

Fig. 5. Gradient index (GRIN) rod lenses of different relative pitch with respect to the rod length: (a, b) quarter-pitch lens, (c) half-pitch lens, (d) one-pitch lens. The arrows indicate the direction of light propagation.

multiple of pitches. The examples in the picture show the propagation in one- and two-pitch rod lenses (Fig. 5c,d). Simulations of GRIN-structures of different index profile functions have been done with the wave propagation method (WPM) [1995_5].

For sufficiently large apertures (i.e. the case of negligible diffraction), analytical rules for an optimum GRIN lens performance can be defined. The index profile for a diffraction-limited waveguiding by a gradient index structure is then simply described by a *hyperbolic secant* function ([1999_1], p. 107):

$$n(r) = n_{max} \cdot \text{sech}(gr) \tag{3}$$

(n_{max} - maximum refractive index, r - radial coordinate, g - $2\pi/P$ gradient constant, P - pitch). For this particular gradient function, the resulting focal length f for a rod lens of the length L is

$$f(g,L) = \frac{1}{n_{max} \cdot g \cdot \sin(gL)} \ . \tag{4}$$

The corresponding working distance W (i.e. the distance between the output facet of the rod lens and the focal plane) is found to be

$$W(g,L) = \frac{1}{n_{max} \cdot g \cdot \tan(gL)} \ . \tag{5}$$

The same properties are also valid for gradient index fibers. However, the propagation in fibers is further influenced by field distortions and losses caused by unavoidable bending.

The spectral and temporal transfer of GRIN rod lenses is essentially limited by dispersion because of the relatively large optical path within the material. Thus, GRIN-lenses are not the optimum choice for beam shaping at ultrashort pulse durations. In other markets like data communication (coupling of waveguides and fibers), multimedia technologies (scanners, copy and fax machines) or the collimation and focusing of high-power diode lasers, however, GRIN-lenses are well established commercial products since many years.

2.4.3. Spherical surface relief lenses

The geometric-optical description of the focusing behavior of surface relief microlenses of homogeneous refractive indices is analogous to the case of macroscopic apertures. The focal length depends on the maximum thickness h_{max} and the radii of curvature ρ_1 and ρ_2. For *thick, convex, perfectly spherical lenses* of rotational symmetry about an optical axis with a refractive index n in vacuum, it can be shown that the focal lengths f and f' measured from the unit planes are (after[1989_3], pp. 161-162)

$$f' = -f = -n\frac{\rho_1\rho_2}{\varDelta}$$

(6)

with

$$\varDelta = (n-1)[n(\rho_1 - \rho_2) - (n-1)t] \ .$$

(7)

The parameter t describes the total lens thickness (distance between the poles of the curved surfaces). It has to be mentioned that eq. (6) is also valid for thin lenses on a substrate of identical refractive index n as schematically depicted in Fig. 6.

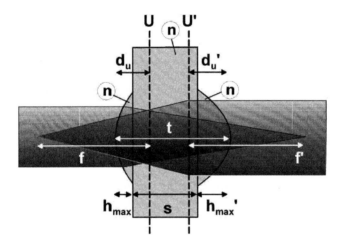

Fig. 6. Thick lens system consisting of two spherically curved areas of the heights h_{max} and h_{max}' rotationally symmetric with respect to the optical axis and a spacer material of the thickness s of identical refractive index n (schematically) This system can be implemented by thin layer structures on a common substrate of identical refractive index.

In this setup (which is exceedingly relevant for thin-film microlenses on transparent substrates), the thickness t is given by the sum of the substrate thickness s and the vertex (sag) heights h_{max} and h_{max}' of the curved, lens-like parts:

$$t = s + h_{max} + h_{max}' \ .$$

(8)

The distances d_u and d_u' between the focal points and the unit planes U and U', are

$$d_u = (n-1)\frac{\rho_1 t}{\varDelta} \ ,$$

(9)

$$d_u' = (n-1)\frac{\rho_2 t}{\Delta} \quad . \tag{10}$$

With eqs. (6), (7), (9) and (10), we obtain for the corresponding absolute values of the focal distances from the lens poles (*working distances*)

$$\left| f - d_u \right| = f \cdot \left[1 - \frac{(n-1)}{n} \cdot \frac{t}{\rho_2} \right] \tag{11}$$

and

$$\left| f' - d_u' \right| = f \cdot \left[1 - \frac{(n-1)}{n} \cdot \frac{t}{\rho_1} \right] \tag{12}$$

In the special case of a biconvex lens of identical radii of curvature ($\rho_1 = \rho_2$), both working distances become equal:

$$\left| f - d_u \right| = \left| f' - d_u' \right| \quad . \tag{13}$$

The radii of curvature of a perfectly *spherical thickness profile* (spherical calotte) can be calculated by

$$\rho = \frac{1}{2} \left[h_{max} + \frac{a^2}{h_{max}} \right] \tag{14}$$

where a is the radius of the free aperture of the lens, and h_{max} the thickness of the calotte (vertex height). The dependence of ρ on h_{max} for selected values of the $d = 2a$ is

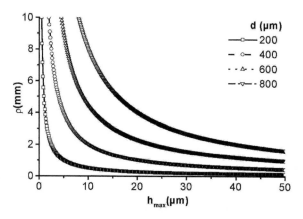

Fig. 7. Radius of curvature of a spherical microlens (calotte) ρ as a function of the thickness (sag height) h_{max} for different lens diameters d (logarithmic ordinate). Please notice the different scales of abscissa and ordinate.

plotted in Fig. 7. In the chosen example, the thickness covers a realistic range for typical lithographically structured microlenses.

The curves show a contrary behavior in two distinct regions. For a fixed diameter, the change of the radius of curvature with increasing h_{max} is slowly decreasing $(dr/dh_{max} \to 0)$. In the opposite direction, the radius of curvature is exponentially rising and increasingly sensitive against small deviations of h_{max} $(dr/dh_{max} \to \infty)$. This leads (in good agreement with the intuitive expectation) to the conclusion that the requirements concerning the shape accuracy are the highest for low-NA structures like thin-film microlenses (see also Chapter 3).

An important special case of a thick lens are sphere-shaped *ball microlenses* which are widely used for laser-to-fiber coupling, fiber-to-fiber coupling, fiber-to-detector coupling and retroreflection (covering reflecting surfaces of traffic signs). With equal radii of curvature $\rho = \rho_1 = \rho_2 = h_{max}/2$ we obtain:

$$f'' = \frac{\rho}{4}\left[\frac{n+1}{n-1}\right] . \tag{15}$$

In the special case of *thin lenses*, i.e. at thickness values small against the radii of curvature ($t \ll \rho_i$), the working distances and focal lengths are nearly identical:

$$\left|f - d_u\right| \approx \left|f' - d_u'\right| \tag{16}$$

$$f \approx f' \tag{17}$$

and the bracket terms in eqs. [11] and [12] vanish. Furthermore, eq. [7] reduces to

$$\Delta = (n-1)[n(\rho_1 - \rho_2)] . \tag{18}$$

Taking these modifications into account, we find the approximation:

$$\left|f - d_u\right| \approx \left|f\right| \approx \frac{\rho_1 \rho_2}{(n-1)[n(\rho_1 - \rho_2)]} \tag{19}$$

and for equal radii of curvature ($\rho_1 = \rho_2$):

$$\left|f - d_u\right| \approx \left|f\right| \approx \frac{1}{2(n-1)} h_{max}\left[1 + \frac{a^2}{h_{max}^2}\right] . \tag{20}$$

To reduce the *spherical and chromatic aberrations*, parabolic microlenses and hybrid structures containing diffractive features can be applied (see paragraph 2.7.). In the case of very broad spectra and ultrashort pulse durations, spectral and geometrical *dispersion* effects are crucial. At very high photon densities like in the case of amplified Ti:sapphire lasers, refractive index changes by *nonlinear effects* further influence the optical system performance significantly. For these reasons, alternative approaches based on thin-film microoptics are currently attracting much interest and will be discussed later in this book.

2.4.4. Fresnel lenses

In Fig. 4e, we find a schematic drawing of a Micro-Fresnel lens (MFL). MFL are radially structured in such a manner that they simulate a continuous phase function by stepwise returning to zero and adapting the phase function up to a maximum phase delay [1987_1, pp. 1-40]. The spatial frequencies of outer parts can be relatively high. Therefore, small and flat Fresnel lenses can act as refractive and diffractive structures simultaneously. The degree of the influence of diffraction is not only determined by the ratio between the surface spatial wavelength and the wavelength of light (see paragraph 2.2.2.), but also by the maximum phase delay. The idea of the MFL mainly takes advantage of the wave properties of light. The phase of a periodical light wave can not be distinguished from its copies with phase delays of integer multiples of the phase corresponding to a complete wave period (2π). In the literature, MFL are therefore also referred to as *diffractive* optical elements (DOE) [1999_1]. Occasionally, this classification causes some confusion so that it is worth to carefully be explained. For monochromatic illumination of a transmissive MFL at the wavelength λ, the refractive index n leads to maximum phase step in each lens zone of

$$t_{max} = \frac{\lambda}{(n-1)} .$$ (21)

The radial zone profile $h(r)$ of a Fresnel lens can approximately be described by the design formula:

$$h(r) \approx \frac{2m\lambda f - r^2}{2f(n-1)}$$ (22)

(r - radial coordinate, m - zone number, n - refractive index, λ - wavelength, f - focal length). Fresnel lenses can also be encoded in low-resolution devices like pixellated SLM [1994_15].

Because of the beam shaping strategy behind (simulation of continuous refractive structures), MFL can be regarded to be refractive elements as well (quite as their macroscopic siblings known from lighthouse lamps or ship lanterns). At least, they are hermaphrodites, of *both refractive and diffractive* nature, as many other structures in microoptics are.

2.4.5. Fabrication of refractive components

To fabricate refractive microlenses, a large number of different methods was developed (see [1999_3], pp. 18-41; [1997_1], pp. 87-143; [1993_13]). *Additive, subtractive and modifying techniques* with respect to the specific procedures of materials treatment and hybrid multistep procedures combining two or more different techniques are known. A well-established method based on a relatively complex process chain is the LIGA technique (lithography with galvano-plating and molding of polymers) [1995_23].

Surface profiles are obtained by grinding and polishing, lithography [2001_21, 2002_24], injection molding (glass, plastic), sculpturing of resist (melting [1993_3, 1997_16], etching, thermal reflow), exposure and squeezing, swelling or structural modification of photosensitive glasses [1985_1, 1986_3, 1986_4, 1990_6, 1990_7, 1991_22, 1992_15, 1994_37, 2004_18], laser heating or direct writing with focused laser [1991_17, 1994_24, 1995_8], light-induced polymerization [1995_6], electron or ion beams, UV laser ablation [1991_21, 1993_12], diamond turning, hydrophobic patterning [2000_22, 2001_6] or sol-gel methods [1997_23, 2002_23], also with lasers [2003_18]. The fabrication of large-NA microlenses by single-step etching followed by mass-transport smoothing [1994_16] and pre-shaping of resist with dry- [1994_17, 1994_18] and wet-etching [1994_25] were reported. Even mushroom-shaped refractive structures were fabricated by resist reflow [1997_19]. Plasma etching and lithographic methods were used to fabricate microlens and antireflection structures in diamond [2003_19]. By a controlled melting of photolithographically structured microcylinders [1989_12], resist microlenses of contact angles varying over a large range [1997_16] with excellent imaging properties [1988_4] can be formed. By filling capillaries with resists followed by UV-curing [1995_24] as well as ink-jet printing techniques [1994_36], arrays of very small structures were obtained. Polymer elements were deposited with computer-aided sputtering methods [1995_25, 1997_20], holographically written in bulk material [1991_18] or obtained by IR-radiation [1996_15].

Other techniques use mask-assisted photochemical materials treatment [1996_13, 1996_14] and shadow-masked metal-organic vapor phase epitaxial growth (MOVPE) [1997_18].

A further promising method is the structured deposition of thin layers, e.g. with laser chemical vapor deposition (LCVD) [1990_2], ion beam deposition [2000_23] or physical vapor deposition techniques which will be described in Chapter 4.

Spatial variations of strain and index as typically found in the case of molding upon heating and cooling and further limitations of a mass production by complicated and time-consuming etching steps can be overcome by contact-less electron-beam polishing that allows to fabricate high-quality microoptics with wide-spread parameters (e.g. aspherical shape functions, arrays of large numbers of elements etc.) [2002_1].

GRIN elements (planar lenses, GRIN-rods, GRIN-type fiber lenses, see paragraph 2.4.2.) are obtained by mask-supported ion-exchange [1995_1, 2002_1, 2001_2] or ion-diffusion [1999_1], also in polymer [1985_2]. Typically, index changes between 0.01 and 0.1 are generated ([1997_1], p. 129). Index modifications can be generated by the irradiation of photosensitive glasses with high-energy photons (see also: [1997_2], pp. 81-89, and [1999_3], pp. 30-38). Distributed index rod lenses in polymer were fabricated by monomer exchange diffusion ([1984_1], pp. 112-122). Bar-, rod- or fiber-shaped GRIN-structures are commercially available.

Refractive microlenses of very small diameters (few wavelengths), hybrid elements containing diffractive features and Micro-Fresnel-lenses (MFL) with small phase steps have to be manufactured with high-resolution technologies. For example, typical necessary structure depths of MFL are in micrometer-range so that lithography is an appropriate technique. Lithographic procedures will be discussed in the frame of diffractive microoptics (see paragraph 2.6.). MFL of low aberrations and high thermal stability were also be obtained by molding procedures [1989_12].

2.5. Reflective microoptics

Reflective microoptics are of ever increasing interest for technical applications as well as fundamental research.

The first reason for this is the need for *large-pixel-number arrays* of steerable micromirrors which are capable for image projection (data projectors, HDTV, digital cinema), measuring techniques (dynamic fringe projection, adaptive wavefront sensing), optical communication (by spatial multiplexing), information processing or lithography (by programming dynamic masks). This can be realized by *dynamic mirror elements*, so-called micro-electro-mechanic systems (MEMS) based on the electrically addressed tilting or rotating of sub-mirrors. To address two-dimensional binary amplitude patterns or to simulate graytone-patterns by local tilt or rotation frequencies, purely reflective elements can be used. Such mirror arrays operate without the crucial polarization problems of liquid-crystal-based spatial light modulators (LC-SLM) and are wasting less of the light energy. The long-term analysis of commercially produced Texas Instruments Digital Micromirror Devices[TM] (DMD[TM]) showed that an extremely reliable performance can be reached [1997_3, 2003_2]. Here, all micromirrors are mounted on tiny hinges and tilted electrostatically at a high speed (kHz frequency range). The grey scale values are realized by a bit-streamed image code, i.e. by the average number of switches per pixel and time unit. The second generation of digital micromirror devices was called Dynamic Light Processing[TM] (DLP[TM]) and operates with up to 2.2 million pixels and dynamic contrast ratio enhancement (DynamicBlack technology). Recently, the DMD Discovery 3000 development platform for high-frequency MEMS-based SLM was introduced [2006_16]. A 0.7" XGA chip (1024 x 768 pixels) works with low voltage differential signaling technology (LVDS), switches at a rate of 200 MHz with a data transfer of 12.8 Gbit/s (16.300 refresh intervals per second) and reaches a contrast of 3000:1.

The second reason for the growing interest in reflective microsystems results from the *necessity to transfer extremely short, spectrally broadband and high-intensity laser pulses* to interaction zones or targets with only minimum distortions by eliminating all specific dispersion or diffraction effects of refractive and diffractive systems. Up to now, a serious application of micromirror arrays to the shaping of high-power laser beams is limited by thermal and spectral problems and also by effective fill factors of available arrays (e.g. 85% in the case of advanced DMD chips). Another aspect is the need for beam shapers in spectral regions of extremely long (FIR) or short (VUV) wavelengths. In the framework of lithographic nanostructuring, UV-capable micromirror devices were developed which have the potential to replace refractive beam shapers (calcium fluoride microlenses). UV-and VUV-micromirrors are of essential importance for micromachining and sample illumination with excimer lasers and higher harmonics of Ti:sapphire lasers. In the case of *frequency-converted and amplified pulses*, the demands to the optics are challenging because of the *combination of short center wavelengths with ultrashort pulse durations and extremely high intensities*. Therefore, further innovative solutions are necessary.

For a large number of applications in beam shaping, *static thin-film micromirrors* are not only a cost-effective alternative to MEMS mirrors but enable to reach an unprecedented performance with respect to the tolerated laser parameters as well as the functionality. Thin-film micromirrors consist of structured metal surfaces, metal layers or dielectric multilayer systems (see Chapters 6 and 7). The specific advantages of such structures (reduced dispersion and absorption, small angles) can be exploited to control the propagation of wavepackets in space and time. *Multilayer* devices allow to realize hybrid refractive-reflective microoptics combining the properties of micromirrors, microlenses and spectral filters with each other. For example, graded reflectance micro-mirror arrays (GRMMA) [1993_5] which have been used to generate beam arrays from Talbot resonators consist of a plurality of elements with Gaussian or super-Gaussian reflectance profiles and an additional phase profile at the design wavelength. In the further course of this book, thin-film microoptical structures and their particular functionality will play an outstanding role.

2.6. Diffractive microoptics

Diffractive optical elements (DOE) take advantage of the diffraction of light waves at substructures comparable to (or even smaller as) the wavelength which has to be minimized in most refractive or reflective systems. The arrangement of substructures can be periodical or aperiodical. The substructures can follow continuous shape functions or be digitized in one or multiple levels (typically in staircase functions) [1990_1, 1992_1, 1993_20, 1993_20]. A particular difficulty in the production of blazed (continuous relief) elements results from the necessary sharpness of phase jumps so that such structures are often well approximated by quantized elements with stairstep-like features. By the choice and composition of materials, certain amplitude and/or phase functions can be obtained. Without trying to present a complete overview on the dazzling array of existing species of DOE and all corresponding theoretical and practical steps of design, fabrication and testing, selected types of particular importance for optical beam shaping will be briefly introduced here.

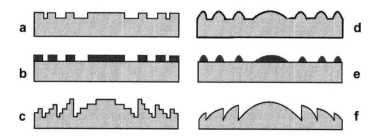

Fig. 8. Selected types of diffractive optical elements (DOE): (a)-(c) quantized (binary) elements, (d)-(f) continuous relief elements (schematically). A two-level binary phase grating (Dammann grating) is shown in (a). Amplitude gratings are drawn in (b) and (e). The structures in (c) and (f) represent a binary multilevel phase grating (kinoform) version of a micro-Fresnel lens and its counterpart with continuous surface relief, respectively.

In Fig. 8, quantized (binary) elements (a-c), and continuous relief elements are schematically compared to each other. Diffractive design can work with phase functions like in the case of the two-level binary phase gratings (Dammann gratings [1971_1]) (Fig. 8a), amplitude gratings (Fig. 8b and e) or combinations of both. More degrees of freedom for the design compared to a two-level profile can be achieved with multi-level kinoform microstructures [1992_1, 1997_24]. These staircase structures approximate the performance of perfect continuous profiles or can even be transformed in those by post-processing. The micro-Fresnel-lenses in Fig. 8c and f represent a binary multilevel phase grating (kinoform) and a corresponding continuous (blazed) structure (see also Fig. 4e).

The design procedures of DOE are typically more complicated as for refractive elements [1993_13, 1995_11, 2003_16, 1994_20, 1995_10]. For the quantization of DOE, different schemes like the Fresnel-zone concept, the superzone concept with a modified phase envelope or the constant pixel concept have been developed ([1999_1], pp. 132-134). Because of the equidistant spatial steps, the constant pixel concept can also be realized dynamically with SLM. The recent progress in fabricating high-resolution SLM with up to a few Megapixels supports this approach so that SLM-DOE might play an increasing role in the near future.

For the theoretical modelling and design of DOE, rigorous vector optics as well as scalar optics are used ([1997_1], pp. 31-52; [1999_1], pp. 129-179; [1999_3], pp. 107-136). The scalar theory neglects the microscopic coupling of electromagnetic field components via the atomic oscillators in inhomogeneous media and takes into account only one scalar transversal field component (e.g. the electric field). For a large number of practical applications, the scalar theory predicts the beam propagation with sufficiently high accuracy. In paraxial approximation (i.e. at small angles), the diffraction can be well described by Kirchhoff's integral. Under near- and far-field conditions, Kirchhoff's integral has to be modified into Fresnel and Fourier transforms corresponding to the cases of Fresnel and Fraunhofer diffraction. Furthermore, the polarization properties of DOE are a further degree of freedom which can be exploited for beam shaping (see: [1997_1], pp. 325-353).

For the fabrication of high-quality DOE, structuring methods of sufficiently high spatial resolution are necessary. Continuous-relief elements can be produced with direct writing methods that are also used for refractive components. Discontinuous, binary structures have to be generated by photolithographic multistep procedures including iterative series of UV- or X-ray lithography and etching ([1997_1], pp. 53-85; [1999_1], 134-139; [1997_2], 23-40; [1991_14]). Arrays of DOE were directly milled into glass with focused ion beams (FIB) [2000_26]. With this technique, rms roughness values of 3 nm were obtained over an area of 2x2 μm^2.

Process steps of a typical photolithographic fabrication technology including subsequent mask exposure of photoresist with UV- or X-ray radiation followed by transfer or replication procedures (see also paragraph 2.8.) are schematically drawn in Fig. 9.

For the optimal control of the chemical etching procedures, the detailed knowledge of the material-specific etching rates is essentially. The practical demands on the accuracy of minimum linewidths and phase step heights are in the range of 5-10%. Typically, total structure depths of few micrometers up to 50 μm are reached. With proton-irradiation-induced surface swelling, even depths of several hundreds of micrometers are possible ([1997_2], pp. 153-176).

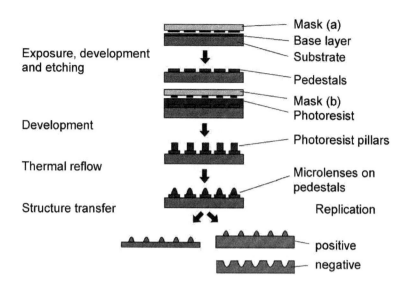

Fig. 9. Multistep lithographic fabrication of microlenses via mask exposure of resist with high-energetic (UV or X-ray) photons (schematically). After structuring pedestals with mask (a) and lenslets with mask (b) and a thermal reflow process, further transfer or replication procedures can be applied (below).

The lateral resolution of available microstructuring technologies is sufficient for realizing sub-micrometer features [1994_21, 1996_11, 2003_24]. With features statistically distributed with respect to size and/or phase step, random phase plates or diffusing panels of tailored angular and polarization characteristics can be implemented.

The spatio-temporal beam shaping of ultrashort pulses or the processing of polychromatic signals with DOE is currently within the focus of interest [2006_19, 2006_20]. For broadband ultrashort pulses, the chromatic aberration of diffractive lenses is relevant which is one order of magnitude larger than for refractive lenses (see: [1999_1], p. 158; [1999_3], pp. 131-133).

2.7. Hybrid microoptics

By integrating the advantageous properties of two or more basic variants of microoptical structures (refractive, diffractive, reflective) in *hybrid components*, new or improved components of significantly extended functionality can be realized.

Diffractive-refractive microlenses are combinations of diffractive structures and refractive microlenses. Fresnel zone lenses with flat blazed substructures (which operate by refractive as well as diffractive principles, see paragraph 2.4.4.) can also be written into thick purely refractive lenses based on smooth surface profiles. This special design is of large practical interest because it enables to further enhance the accessible numerical aperture and to compensate for chromatic and spherical aberrations as well as thermal expansion effects ([1997_1], pp. 259-292). With hybrid athermal lenses (so-called "athermats") consisting of only a single material, significantly improved thermal behavior is obtained. In Fig. 10, the blazed surface profile of a hybrid diffractive-refractive lens of about 50 µm height detected with an electron microscope is shown. The structure was lithographically written in photoresist (from [1996_1]).

By implementing an achromatic hybrid lens design, the performance of beam array generators can be optimized. [1999_19].

The potential of hybrid concepts for applications with spectral selectivity was not only demonstrated by an aberration correction but also, vice versa, by exploiting the dispersion for hyperspectral sensor systems ([1994_1], pp. 329-359). First of all, however, thin-film approaches like structured multilayers [2001_1], pseudo-reflective metal-dielectric compound optics [2003_1] or combinations of diffractive gratings with thin refractive layers [2006_8] promise advanced hybrid microoptical design solutions which are rather difficult to mimic with other available technologies (see Chapter 3). Particular features of multilayer microoptics will be addressed in the context of laser mode selection, ultrashort-pulse laser beam shaping and pulse diagnostics (see Chapters 6 and 7).

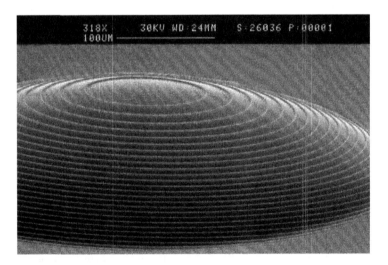

Fig. 10. Hybrid photoresist microlens consisting of a thick refractive surface-relief with additional blazed substructure (total height 50 micrometer, measured with electron microscope; courtesy of Institute of Applied Physics, Friedrich-Schiller-University Jena, Germany [1996_1]).

2.8. Replication and structure transfer

Microoptical elements can be replicated from master components to reduce the fabrication costs and to obtain a large number of structures in a short time, and they can be transferred into substrates by chemical or physical procedures to end up with stable, monolithic structures. Thus, any problems of internal stress, Fresnel reflection at interfaces or thermal expansion can be successfully eliminated. The price one has to pay for both replication and transfer is a certain loss in the shape accuracy and surface roughness (depending on materials and process parameters).

Well-known replication methods are resin casting, hot embossing, injection molding or UV-embossing ([1997_1], pp. 153-177). All these methods are characterized by specific advantages as well as specific disadvantages with respect to the process speed, reachable structure depth, resolution and aspect ratio.

A transfer of layer structures like melted photoresist [1999_20] or oxidic thin film microlenses [1997_5, 2004_1, 2005_9] into substrates can be achieved by reactive ion etching. The following two special cases show different advantages (Fig. 11):

(a) layer and substrate consist of different materials

(b) layer and substrate consist of identical materials

In case (a), the etching selectivity can be used to control the final structure depth [1999_20]. This enables for a proportional transfer of extremely flat structures (schematically in Fig. 11a) which are needed for low-NA microlenses (e.g. for VUV excimer laser beam array shaping [2001_3, 2004_1, 2005_9, 2006_6]). In case (b), an etching selectivity of 1:1 can be used to identically transfer structures (schematically in Fig. 11b) and to reduce problems by different thermal expansion coefficients or chemical properties. Furthermore, the optical parameters of composite structures of

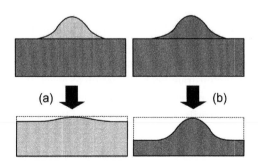

Fig. 11. Selective structure transfer: Etching of surface microstructures from a thin layer into a substrate for (a) materials of different etching ratios and (b) identical materials of layer and substrate (above: initial composite structures, below: final monolithic structures; schematically).

identical materials can be tested already before the transfer process. For example, the interferometric characterization of composites of two different materials is often accomplished by parasitic reflections or backside signals so that auxiliary reflective layers are necessary. This problem can be avoided to a large extent in case (b).

Replication and transfer methods can also be combined as it was demonstrated, e.g., for excimer laser beam homogenizer [1997_27]. Here, master structures were fabricated by direct writing e-beam lithography, replicated by hot-embossing and transferred into UV-grade fused silica by reactive ion etching (RIE).

2.9. Specific properties of array structures

2.9.1. Types and general features of microoptical array structures

Arrayed arrangements of multiple microoptical elements are an attractive design approach in nature (complex eyes) as well in technical optical systems ([1988_4, 1991_12, 1991_15, pp. 173-215, 1999_3, 2002_32]). Collimation, focusing, coupling or filtering tasks in systems with array-shaped sources, multiple fibers or waveguides, matrix detectors or array masks can be advantageously solved with adapted microlens arrays. For example, large numbers of holes can be drilled into hard wings of airplanes (to improve the aerodynamic properties) in a cost-effective way by simultaneously generating many focal spots from a single excimer laser beam [1992_13]. Microlens arrays are implemented in optical relay systems [2003_25], SLM [1994_26], and filtering stages of optical neural networks [1990_3, 1990_4, 1991_23].

Assembly, mounting and adjustment in multichannel systems are essentially simplified by using integrated 1xN- or MxN-type arrays instead of handling many individual elements [2001_23]. The splitting of beams into subbeams with subapertures enables for high-speed data transfer, large-field-of-view imaging and simultaneous materials processing by decoupled information or energy transfer in separated channels. The number of channels can reach from 1x2 in simple switches up to the range of 10^7 for high-resolution image detectors, diffusors, displays or advanced spatial light modulators. As we will demonstrate later, the option to generate two-dimensional maps of spectral or temporal beam parameters and the parallelism of a number of excitation channels are also extraordinarily interesting for the diagnostics and applications of ultrashort wavepackets.

With respect to the *geometrical arrangement*, there exist *regular* (e.g. orthogonal, hexagonal, linear, circular etc.) arrays and *irregular* multielement structures (which, per definitionem, do not belong to the class of arrays). Furthermore, one has to differ between *uniform arrays* (in ideal case arrays consisting of identical elements), and *non-uniform arrays* (consisting of elements with spatially varying or addressable parameters like focal length or, more general, local phase function). In special cases, non-uniform arrays can be spatially modulated with defined *envelope functions* of phase and/or amplitudes [1997_5] as also shown for mode-selective resonator structures [1994_2, 1994_3], optical testing [1997_4] and the temporal diagnostics of ultrashort signals [P7].

Multichannel imaging systems based on microlens arrays can be used for the contactless photolithograpic production of microstructures on large print areas [1996_17, P15]. The method can be applied to relatively thick photoresist layers (limited by the depth of focus which is typically > 50 µm. The achievable uniformity of arrays depends on the fabrication technology, substrate quality, and number, size and shape of the elements. Typically, commercial microoptical arrays reach uniformity values between 1 and 3 %.

Multichannel interferometry and microscopy can be realized with microlens and microaxicon arrays. Array-based confocal microscopy enables for a fast contactless surface profiling [1994_13, 1996_12].

In contrast to single beams, the propagation of multiple beams has to be described by particular beam propagation constants, see e.g. [1998_8].

2.9.2. Fill factor, efficiency and symmetry

For the efficiency of light transfer through microoptical arrays, the *fill factor* is of essential importance. The most imaging as well as non-imaging systems require a maximum efficiency, i.e. the ratio of the light energy that reaches the target zone (image detector, fiber array, solar cell, laser medium, resist etc.) to the input energy. The losses of an array of identical microlenses result from two different sources: (a) the losses of the single elements (absorption, Fresnel reflection, diffraction, aberrations etc.) and (b) the losses induced by the *geometrical fill factor*. The geometrical fill factor η is defined as the ratio of the area filled by the array elements (A_{ML}) to the total area of the array (A_0):

$$\eta = \frac{A_{ML}}{A_0} \tag{23}$$

The maximum possible geometrical fill factor η_{max} is obtained for the densest package of elements and depends on the shape and the arrangement of the elements. Values for selected types of arrays corresponding to the case of the densest package are compared in Fig. 12.

The fill factors of the first two arrays in the graph reach 1 in ideal case. In praxi, however, square-shaped microlenses typically suffer from significant shape deviations near the corners which reduce the efficiency. The exact fill factor of arrays with dead

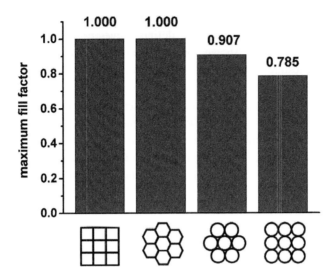

Fig. 12. Maximum possible geometrical fill factors (densest package) for different types of arrays and shapes of elements (from left to right: orthogonal array of square-shaped elements, hexagonal array of hexagonal-shaped elements, hexagonal array of circularly shaped elements and orthogonal array of circularly shaped elements).

space like in the case of circular or elliptical elements or distances between the elements depends on the size and number of elements. As it can easily be shown, the fill factor of a hexagonal array of circular elements approximates in the limiting case of sufficiently much elements the value

$$\eta_{max} = \frac{\pi}{6}\sqrt{3} \ . \tag{24}$$

For an orthogonal array of circular elements, the maximum value is

$$\eta_{max} = \frac{\pi}{4} \ . \tag{25}$$

The array structures differ not only in fill factor but also in their *symmetry properties*. This is reflected by different diffraction patterns of orthogonal and hexagonal arrays generated by their specific spatial frequency contents. For applications like phase-coupling by self-imaging effects (see paragraph 2.9.4.) or channel separation, fill factor and symmetry have to be taken into account. Optical processors, free-space relay systems and the design and performance of wavefront sensing devices (see paragraph 2.9.6.) are closely connected to both parameters.

For all three basic modes of microstructuring (subtracting, modifying or adding material), appropriate methods to *fabricate arrays of high fill factors* were developed. A relatively simple way is to subsequently write linear arrays of cylindrical elements (e.g. microlenses or gratings) in one and the same material but at different orientations (e.g. 2x for orthogonal arrays, 3x for hexagonal arrays etc.; see also Chapter 4). The *method of crossed interaction zones* was first developed for writing orthogonal and hexagonal microlens arrays via holographic illumination and index modification of glass and polymers [1985_3, 1991_18]. It was also applied to point-by-point or line-by-line *crossed scan* etching and ablation procedures [1994_29] and crossed deposition techniques with shadow masks [1996_4, 1999_7] (see Chapter 4). If the crossed zones differ in depth of diameter, arrays of *anamorphic* microlenses can be fabricated [P3], ([1997_5], pp. 169-179). Other methods work with single-step procedures [1997_20].

In a more generalized case, spatially *crossed interaction zones* (which can be patterned in a complex manner to create 3D arrangements of structures) are superimposed.

2.9.3. Spatial frequencies and Fresnel numbers

The spatial phase and/or amplitude profile of arrays is characterized by additional spatial frequencies causing particular diffraction patterns. In order to describe those diffraction phenomena, the simple picture of a Fresnel number has to be generalized. Each spatial frequency component

$$v_i = \frac{1}{\Lambda_i} \tag{26}$$

within the spatial frequency spectrum of an array structure of spatial period Λ_i corresponds to a certain Fresnel number

$$F_i(z) = \frac{1}{4v_i^2 \lambda z} = \frac{\Lambda_i^2}{4\lambda z} \tag{27}$$

for a given wavelength λ and distance z. A simple way to get an idea about the origin of the particular diffraction pattern of an array is to search for the minimum and maximum relevant spatial frequencies. The minimum spatial frequency $v_{min} = 1/\Lambda_{max}$ is given by the maximum diameter of the array which can be expressed by the period p, the diameter d of a single element and the number N of elements fitting within d:

$$v_{min} = \frac{1}{\Lambda_{max}} = \frac{1}{Np+d} \tag{28}$$

The maximum spatial frequency corresponds to the array period p (center-to-center distance between nearest neighbors):

$$v_{max} = \frac{1}{\Lambda_{min}} = \frac{1}{p} . \tag{29}$$

With eqs. (28) and (29) we obtain the maximum and minimum Fresnel numbers:

$$F_{max}(z) = \frac{1}{4v_{min}^2 \lambda z} = \frac{\Lambda_{max}^2}{4\lambda z} = \frac{(Np+d)^2}{4\lambda z} \tag{30}$$

and

$$F_{min}(z) = \frac{1}{4v_{max}^2 \lambda z} = \frac{\Lambda_{min}^2}{4\lambda z} = \frac{p^2}{4\lambda z} . \tag{31}$$

In the limiting case of $p = d$ (100% fill factor), eq. (31) would be identical to the case of a single diffracting aperture (compare eq. (1)). The local structures (high-frequency part of the spatial frequency spectrum) are responsible for the envelope spatial frequencies of the far-field patterns. Vice versa, the global array structure (low-frequency part of the spatial frequency spectrum) is responsible for the local spatial frequencies of the far-fields. Without going into the details, it should be mentioned that this optical complementarity is virtually a textbook-example for the correspondence between optical wave diffraction and mathematical Fourier transform.

2.9.4. Cross-talk and self-imaging effects

Depending on the element type, subaperture diameters, element distances, or spectral and spatio-temporal properties of the radiation to be transformed, array architectures show further specific effects in comparison to single channel systems. A very important phenomena is the *cross-talk* of neighboring channels which can be caused by refraction, diffraction and scattering (in free-space systems) or dispersion, waveguiding and parasitic reflection (in waveguiding systems and within refractive lenslet arrays). Cross-talk decreases the signal-to-noise ratio and causes errors in multichannel data transfer, whereas it also can be exploited to coherently couple arrays of light emitters or to reshape and homogenize beams.

Specific applications are enabled by so-called *self-imaging effects* [2001_26] which result from the self-reconstruction of periodical components of wave fields [1989_13, 1998_15] at discrete distances (Talbot effect [1836_1], Montgomery effect [1967_1, 1968_1], fractional or fractal Talbot effect [1965_1, 1996_18, 2004_23], and fractional Montgomery effect [2005_26]; see also paragraph 7.6). The *Talbot-effect* is related to the recovery of a periodic phase and/or amplitude pattern in space, time and spectrum. Assuming the paraxial case, perpendicular illumination and a monochromatic source, the initial *phase and/or amplitude* pattern is reproduced in good approximation by constructive interference at certain distances z_T, at the so-called *Talbot distances*:

$$z_T = 2m \frac{p^2}{\lambda} = \frac{2m}{\lambda v_P^2} \tag{32}$$

(m - integer number or integer fraction, p - array period, v_p - spatial frequency corresponding to the array period p, λ - wavelength). The geometrical conditions for the self-imaging by free-space Talbot effect are schematically drawn in Fig. 13 for the phase replication behind a microlens array.

In the literature a controversial discussion about the exact interpretation of the Talbot effect was held (see, e.g., [1992_10], [1993_16], [1993_17]). In particular, the effect was described as self-images by Fourier optics, multiple-slit diffraction patterns in Fresnel and Fraunhofer domain and a superposition of nondiffracting beams or modes. In a more recent paper, an explanation by applying the number theory to rays was presented [2001_27]. In focusing experiments with microlens arrays of high spatial frequencies, the Talbot effect can easily lead to confusions because of a mix up of Talbot planes with the focal plane. Integer fractions of m correspond to the *fractional Talbot planes*. A self-imaging of only the amplitude pattern with transversally *alternating phase* is obtained in other planes located between the Talbot planes.

Among the most popular applications of self-imaging effects are several approaches for a coherent coupling of arrayed laser emitters by generating supermodes (see, e.g., [1986_1, 1988_6, 1988_7, 1989_14]). However, a detailed analysis of the Talbot effect shows that the matter is accomplished by deviations of amplitude and phase patterns in the Talbot planes compared from the initial distributions [1993_18]. Therefore, even perfect Talbot resonators without additional beam correction work with significant losses.

High-efficiency Talbot array illuminators which were first proposed by A. W. Lohmann [1988_5], [1993_15] exploit the Talbot effect at phase gratings [1997_28] or microlens arrays [1994_28] and can be applied to free-space interconnects [1990_5], optical neural networks [1990_4], or for single-shot measurements of laser damage thresholds [2000_27].

In his original experiment [1836_1], H. F. Talbot demonstrated the self-imaging phenomenon with a point-like white light source and observed a color separation. The spectral selectivity of the *polychromatic* Talbot effect can be used to build up compact spectrometers [2003_22]. Talbot [2001_25, 2006_17] and Montgomery interferometers [2004_24] can serve as time filters for the temporal multiplexing of optical signals. The application of the Talbot effect in linear mode [2002_28, 2002_31] and with nonlinear frequency conversion [P5] was proposed. With femtosecond Talbot experiments, a

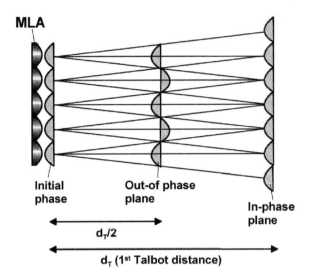

Fig. 13. Phase self-imaging of array components: Replication of regular phase patterns by the Talbot effect (MLA - microlens array, d_T - Talbot distance). The microlens array is illuminated from the left side by a plane wave. The initial amplitude pattern which is not shown here is also replicated at the out-of-phase planes. Both phase and amplitude are approximately replicated at the Talbot planes (here: first Talbot plane). At the outermost elements, the amplitude is reduced by the missing contributions by next neighbors (schematically).

coherence mapping of few-cycle wavepackets was performed in linear regime [2001_9, 2001_10, 2002_10]. However, it has to be pointed out that linear methods are not sufficient to completely characterize the temporal behavior without time-nonstationary elements [1996_19, 2002_29].

It should be mentioned that self-imaging effects have also been studied at extremely small wavelengths. For example, the Talbot effect from solid state materials under X-ray illumination gives new insight in the dynamics of atom lattices. It has to be expected that particular knowledge on optical self-imaging can (after necessary modifications) be applied to the X-ray domain, e.g. for atom-scale lithographic structuring, molecular engineering or data storage.

2.9.5. Wavefront detection with array components

Arrays of microoptical components can be used to reconstruct the curvature profile of a wavefront by simultaneously detecting the local wavefront tilt at discrete positions. A highly precise wavefront diagnostics enables for correcting wavefront data of star light in astronomy ([1992_11, 1992_12, 1994_30, pp. 121-132, 1995_17]) or laser materials processing. To analyze the wavefront tilt, any approach can be used which locally transforms the angular information into a spatial one. This is possible by using arrays of holes (Hartmann sensor) or microlenses (Shack-Hartmann sensor) [1995_16] to generate subbeams the direction of which is monitored by the transversal shift of the positions of Airy disks or focal points. Alternative types of wavefront sensors can be realized with arrays of micro-interferometers (e.g. space-variant Fabry-Pérot etalons) where deviations of fringes from the radial symmetry indicate the local wavefront tilt as well. A schematic comparison of three of the approaches is shown at Fig. 14.

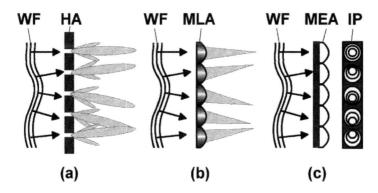

Fig. 14. Wavefront sensing with different types of multichannel detectors: (a) Hartmann sensor detecting the local tilt of the wavefront (WF) with a hole array (HA), (b) Shack-Hartmann sensor detecting the wavefront tilt by splitting the beam with a microlens array (MLA) into an array of focused subbeams, and (c) micro-etalon array (MEA) detecting the wavefront by generating angular dependent interference fringe patterns (IP) (schematically).

The drawbacks of the Hartmann-type device are the low fill factor and the diffraction at the sharp edges of the pin holes. Shack-Hartmann-sensors can be operated in scanning mode (moving by one or a few pixels) to improve the spatial resolution [1999_17]. Unwanted cross-talk by diffraction ripples can be reduced by using *apodized* elements (i.e. elements with smoothed pupil functions) [1999_18] or by adaptive masking of elements with spatial light modulators. Recently, further improvements of the Shack-Hartmann sensor principle have been proposed [2004_5, 2004_13, 2004-14, 2005_6, 2005_7, 2005_8, P12]. Here, the microlenses are replaced by axicons (conical lenses or mirrors) generating extended focal zones (so-called nondiffracting beams [1987_8, 1987_9, 1988_1], see Chapter 6) [1954_1]. Because of the large depth of focus, low aberrations and other advantageous properties of such beams, axicon-based Shack-Hartmann sensors might become key elements for the analysis of laser beams at extreme spatio-temporal and angular parameters. The principle enables to realize reflective setups so that it is very attractive for ultrashort-pulse characterization (see Chapter 7).

Information on wavefront curvatures can also be derived from analyzing the distortions of the above described Talbot effect. This method was recently adapted for X-ray wavefront measurements [2005_27] which contribute to establish next generation lithographic technologies. Because of the reduced diffraction at short wavelengths and the material-free propagation through pinholes, also Hartmann sensors experience a renaissance in X-ray applications.

2.9.6. Gabor superlens

A further array-specific phenomenon was invented by Denis Gabor in 1940 [P14]. The *Gabor superlens* is a microoptical system formed by a pair of microlens arrays which slightly differ in their array periods. If the resulting long-period beat frequency of the spatial frequencies is calibrated in a proper way, the local pairs of microlenses act as beam steering elements generating an array of converging subbeams (see Fig. 15).

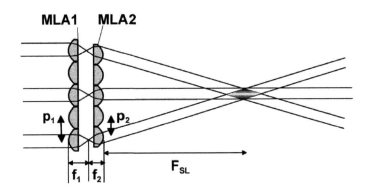

Fig. 15. Gabor superlens effect as a superposition of subbeams generated by two arrays of microlenses of slightly different spatial period (schematically, after [1997_33]).

Thus, the whole system behaves like a big, facetted lens, ("superlens") [1999_21]. The effect can also be observed if the arrays are slightly tilted with respect to each other). The Gabor superlens is known from applications in advertising labels and postcards where arrays of small pictures are transformed to a single, magnified image with a pseudo-3D-effect. The effect has the potential to be extended to applications in the field of coupling, laser mode-selection and wavefront sensing if the inevitable losses can be tolerated. In combination with adaptive liquid-crystal microlenses or spatial light modulators, the Gabor effect could be realized in dynamic mode.

Similar, closely related *Moiré effects* can be observed in the case of only a single microlens array (a) if the array is illuminated by a quasi plane wave and placed at a distance near the focal length of the single microlenses from a screen thus imaging the focal spot array under the influence of a slight tilt or wavefront curvature and (b) if a microlens array images an array of emitters of slightly different pitch (e.g. the pixels of an LCD monitor). The Gabor effect and the mentioned related effects can be used to measure the focal length and to evaluate the uniformity of microlens arrays.

2.10. Stacked and planar microoptics

Beside two-dimensional array structures, the extension to the third spatial dimension [1993_13] and the folding of optical pathways in flat system architectures are of increasing interest to realize compact systems. For converting the numerical aperture in one or two spatial directions, multiple microoptical building blocks were integrated in *stacks* ([1984_1], pp. 201-206; [1999_1], 211-213; [1997_1], pp. 199-221). Stacked systems are used, e.g., to realize crossover interconnects [1995_13].

A further approach to a higher degree of integration came with the concept of *planar optics* (see: [1999_1], pp. 213-215; [1997_1], pp. 179-198). In planar-optical arrangements, microoptical elements are integrated side by side in planes on the surfaces of common substrates. The compactness is enhanced by folding the light path within the system. Therefore, aberrations must be minimized for oblique angles of incidence, and reflective and off-axis design are necessary. For example, optical processors for fractional correlation were built up in planar-optical arrangement [1997_30].

2.11. Problems and trends

Currently we observe numerous trends and overlaps in the fields of micro- and nanooptics which are driven by the progress in fabrication technologies (higher resolution, higher precision, parallelism, new materials, shorter wavelengths), new demands from the markets (multimedia technologies, microbiology, medicine, homeland security, communication) and challenges from fundamental research (astronomy, laser beam shaping, atom manipulation, Bose-Einstein condensates, confocal microscopy, quantum information etc.).

Particularly remarkable trends are:

- the development of *adaptive microoptical components* like liquid crystal microlenses of steerable focus length [1997_31] or such of hydraulically controlled shape
- programming of DOE functions into *SLM* [1994_35, 2001_28]
- micro- and nanostructuring of *materials* with special thermal or spectral characteristics (e.g. diamond [2003_19], silicon, calcium fluoride)
- *solid immersion* microoptics [2001_29]
- *photonic crystal structures* [2003_4]
- high-resolution *near-field optics* and the optics of evanescent fields
- passive and adaptive *plasmonic devices*
- metamaterials with *negative refractive index* in visible range
- integration of microoptics and MEMS / MOEMS [1994_8, 1994_9]
- development of *nano-electro-mechanical and nanooptical-electro-mechanical systems* (NEMS, NOEMS) [2002_33, 2005_29]
- transfer of fabrication technologies to and integration of microoptical elements into *microfluidic* systems [2006_21]
- *axial beam shaping* [1998_3]
- *polarization and orbital momentum* engineering for superresolution and particle handling [1999_12, 2002_13, 2002_14, 2002_22, 2006_11, 2006_12]
- diffraction control by *apodization* [2006_8]
- *polarization self-imaging* [2002_30]
- beam shaping of higher complexity (tailoring of the *Wigner function*)
- sophisticated and adaptive *light diffusing and distributing systems* including miniaturized non-imaging elements (e.g. for car components, windows, mobile phones, solar reactors etc.)
- *quantum structures* (quantum dots, quantum wells, quantum wires)
- *integration of lenses and prisms* in monolithic semiconductor structures like *vertical-cavity surface-emitting lasers* (VCSEL)
- *multilayer microoptics* [2001_1]
- *sculptured layers* [2005_1]
- *X-ray microoptics* [2002_34]

It should also be mentioned that the further improvement of *techniques for mounting and adjusting* microoptical systems [1995_12] is a key problem for their implementation in real-world devices, in particular with respect to efficiency and reliability.

The study of *biological micro- and nanostructures* with respect to their optical properties will lead to novel technical solutions (*bionics*) (e.g. artificial retina sensors, facet eyes and telescope compound eyes [1993_19, 1993_21, 1994_31, 1996_20,

1996_23, 1995_2, 2004_25, 2004_26, 2005_28]). The progress in structural and functional complexity has to be supported by new and improved high-resolution characterization techniques and test procedures for industrial standards (further development of ISO, e.g. for arrays [1991_2], compare [1997_21]).

New applications of commercially available high-power laser systems require the extension of microoptics to new frontiers like

- short wavelengths (UV, VUV, X-ray)
- ultrashort pulse durations (femtoseconds, attoseconds)
- ultrabroadband spectral profiles (up to octave spanning and more)
- extremely high power (Petawatt lasers)
- nonparaxial fields (high-power laser diode arrays)

However, particular limitations arise for the propagation of ultrafast signals through microoptical systems. Typically, such systems act as a bandpass reducing the bandwidth and enhancing the pulse duration so that unconventional components have to be designed. The spectral, angular and travel time filtering of wavepackets by diffraction, material dispersion and geometrical dispersion can be minimized by using *small angles, low thicknesses or reflective devices*. As it will be shown later, the optimum transfer functions can be well approximated by *thin-film* microoptical components. The recent progress in just this direction belongs to the most exciting things that happen in optics at the moment and is a major objective of this book.

Chapter 3

THIN-FILM OPTICS

3.1. The concept of thin-film optics

Parallel to the miniaturization of optical and electronic devices, high-precision techniques were also developed for the deposition of thin dielectric and metal layers. By applying adapted simulation algorithms, it became possible to accurately predict spectral transfer functions for even complicated stacks of numerous dielectric layers. Thus, even highly functional components like broadband antireflection coatings and mirrors for few-cycle femtosecond lasers, X-ray reflectors, narrowband filters or gradient mirrors (to mention only a few) can be well designed and fabricated.

The concept of thin-film optics comprises at first all the components based on layers or layer systems and secondly all the specific optical laws resulting from low thickness (compared to "classical" optical hardware) and/or structure and composition (e.g. in the case of layer stacks). Until now, the principal reasons for the application of optical coatings were the "*modification of reflectance and to some extent transmittance and occasionally absorptance*" (H. A. McLeod, in [1995_26, p.1]). Therefore, the synthesis of microoptical and thin-film optical concepts has still the charm of a new and unexploited field which benefits from the mutual exchange of knowledge. Concerning the numerous details of deposition techniques, materials and simulation software for coating design, one can refer to a large number of excellent publications (see. e.g., [1970_1, 1987_5, 1987_6, 1988_14, 1995_26, 1996_16, 2002_27, 2002_36, 2003_26]). In view of the application to spatio-temporal beam shaping, we will mainly concentrate on the fundamentals of the light propagation through thin optical structures where *spectral interference effects* play a key role. In the following sections, the response of mono- and polychromatic light fields to uniform and nonuniform single and multiple layers will be described with respect to the "classical" applications as well as typical microoptical configurations.

3.2. Single transparent layer on substrates

In the most simple case, a thick transparent substrate of a refractive index n_S and an absorption coefficient α_S is coated with a single, thin optical layer of a refractive index

n_L and an absorption coefficient α_L which has a constant thickness h over all spatial positions (x,y) on the substrate. Then, the light propagation through the layer is mainly influenced by the Fresnel reflection at surface and interface resulting. The resulting Fabry-Pérot etalon resonances depend on the spectral bandwidth, angular distribution, absorption, dispersion and scattering. The determination of optical constants by the analysis of fringes of highly absorbing and scattering semiconductor layers was demonstrated already in the 1960s [1968_2]. For transmittive microoptical applications, vice versa, layers of low absorption and scattering losses are of particular interest. Several special cases of such layers will be considered now.

3.2.1. Uniform Fabry-Pérot etalon at monochromatic illumination

For a *single layer of uniform thickness distribution* and constant positive refractive index n_L with negligible absorption on a substrate of different refractive index n_S, the Fresnel reflection of plays the role of etalon mirrors generating multiple beam interference (Fig. 16).

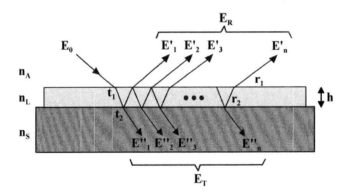

Fig. 16. Multiple beam interference in a uniform dielectric thin layer of the thickness h and refractive index n_L without absorption on a substrate of refractive index n_S (schematically). The accumulated reflected and transmitted partial field amplitudes E'_i and E''_i deliver the total reflected and transmitted fields E_R and E_T. The amplitude reflectivity and transmittivity coefficients at air-layer and layer-substrate interfaces are referred to as r_1, r_2, t_1, t_2, respectively.

Here, all partial field components have to be summed up to calculate the relevant transfer functions. As the sum for the transmitted field at a wavelength λ at normal incidence of a plane wave we obtain [2001_31, pp. 77-81]:

$$E_T(\lambda) = E_0 t_1 t_2 \sum_{m=0}^{\infty} (r_1 r_2)^{2m} e^{im\phi}$$ (33)

(E_0 - initial field amplitude, r_1, r_2 - amplitude reflectivity coefficients). The phase term

$$\varphi = 2n_L \cdot h \cdot \frac{2\pi}{\lambda} = 4n_L \cdot h \cdot \frac{\pi}{\lambda}$$ (34)

is proportional to the optical path length (optical thickness) for a single roundtrip within the layer:

$$L_{RT} = 2n_L h .$$ (35)

In the case of oblique illumination at an angle of incidence φ, the formulas in eqs. (34) and (35) have to be multiplied with the correcting factor $\cos\varphi$.

The resulting total transmitted field $E_T(\lambda)$ is found to be:

$$E_T(\lambda) = E_0 t_1 t_2 \frac{1}{1 - r_1 r_2 e^{i\phi}}$$ (36)

At normal incidence, we obtain the refractive indices of the interfaces air-layer and layer-substrate from Snell's law as

$$n_{AL} = \frac{n_L}{n_A} = n_L$$ (37)

(with a refractive index of air $n_L = 1$) and

$$n_{LS} = \frac{n_S}{n_L} .$$ (38)

The corresponding transmission and reflection coefficients for the electrical field amplitude at the layer-substrate and layer-air interfaces with respect to parallel and perpendicular polarization are obtained from the Fresnel formulas [1989_3, p. 41]:

$$t_{1p} = t_{1s} = \frac{2}{n_L + 1}$$ (39)

$$r_{1p} = \frac{n_L - 1}{n_L + 1}$$ (40)

$$r_{1s} = -\frac{n_L - 1}{n_L + 1}$$ (41)

$$t_{2p} = t_{2s} = \frac{2}{n_{LS}+1} = \frac{2}{\dfrac{n_S}{n_L}+1} = \frac{2n_L}{n_S+1} \tag{42}$$

$$r_{2p} = \frac{n_{LS}-1}{n_{LS}+1} = \frac{\dfrac{n_S}{n_L}-1}{\dfrac{n_S}{n_L}+1} = \frac{n_S-n_L}{n_S+n_L} \tag{43}$$

$$r_{2S} = -\frac{n_{LS}-1}{n_{LS}+1} = -\frac{\dfrac{n_S}{n_L}-1}{\dfrac{n_S}{n_L}+1} = -\frac{n_S-n_L}{n_S+n_L} \tag{44}$$

The negative sign of the amplitude reflectivity coefficient in formula (44) indicates a phase jump by π. From eq. (36). The *transmitted intensity* I_T can be expressed in complex form:

$$I_T = I_0 \frac{(t_1 t_2)^2}{(1-r_1 r_2 e^{i\varphi})^2} \tag{45}$$

or in the real notation [2002_39, pp. 108-118]:

$$I_T = I_0 \frac{(1-r_1^2)(1-r_2^2)}{1+(r_1 r_2)^2 - 2r_1 r_2 \cos(\varphi)} . \tag{46}$$

In eq. (46) the negative sign of eq. (44) has to be taken into account. Because of the conservation law for the energy, the *reflected intensity* (in absence of absorption and scattering losses, and under monochromatic conditions) is:

$$I_R = I_0 - I_T . \tag{47}$$

Figure 17 shows the calculated intensity transmitted though a uniform fused silica layer on a plane polymer substrate (refractive indices $n_L = 1.46$, $n_S = 1.585$, wavelength $\lambda = 800$ nm) as a function of the parameter $h \cdot n_L / \lambda$ (optical thickness normalized to the wavelength). The simulation was performed with a MATHCAD® program. The minima and maxima in the transmission curve in Fig. 17 correspond to the etalon resonances.

The *transmission contrast* of this layer etalon given by the maxima I_{Tmax} and minima I_{Tmin} of the transmitted intensity

$$C_T = \frac{I_{T\max} - I_{T\min}}{I_{T\max} + I_{T\min}} \tag{48}$$

was found to be about 0.013. In general, a low index step between a transparent layer and a transparent substrate yields a low intensity modulation. To avoid parasitic interference fringes or feedback in thin-film structures, both materials should be identical. However, in particular cases it can be just interesting to generate interference fringes (e.g. for coherence mapping or wavefront sensing, see paragraphs 6.4 and 7.2).

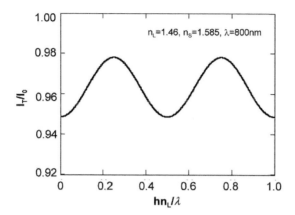

Fig. 17. Thickness dependence of the multiple beam interference in a uniform thin dielectric layer (fused silica on polycarbonate substrate, $n_L = 1.46$, $n_S = 1.585$, monochromatic plane wave illumination at a wavelength of $\lambda = 800$ nm). The dimensionless thickness parameter $h \cdot n_L/\lambda$ was varied between 0 and 1.

Further parameters which are widely used to characterize the performance of etalons are the *spectral bandwidth*, the *free spectral range*, and the *finesse*. The spectral bandwidth of etalons is commonly described by the spectral interval $\Delta\lambda_{FWHM}$ corresponding to the low frequency and high frequency full width at half maximum (FWHM) values of the transmitted intensity as a function of wavelength [2001_31, pp. 77-81]:

$$\Delta\lambda_{FWHM} = \frac{4\pi L_{RT}}{\arcsin\left(\frac{1-r_1 r_2}{2\sqrt{r_1 r_2}}\right)} = \frac{\varphi \cdot \lambda}{\arcsin\left(\frac{1-r_1 r_2}{2\sqrt{r_1 r_2}}\right)} . \tag{49}$$

The FWHM is proportional to the root of the second order moment of the spectral response function of the system. The free spectral range is the distance of the intensity maxima and corresponds to the product of FWHM and finesse F:

$$\Delta\lambda_{FSR} = \Delta\lambda_{FWHM} \cdot F \tag{50}$$

where
$$F = \frac{\pi\sqrt{r_1 r_2}}{1 - r_1 r_2} \ .$$
(51)

However, as we can easily learn from the weakly modulated curve in Fig. 17, these parameters can only be applied to etalons of sufficiently high reflectivity (i.e. with high contrast). If the modulation depth of the spectrally dependent transmitted intensity is below 0.5, the FWHM is *undefined* and eqs. (49) and (50) lose their physical sense, although a free spectral range can be measured experimentally. For a more appropriate description of the spectral behavior of weakly modulated intensity transmission, adapted mathematical procedures have to be applied (higher order statistical moments [2006_22], separation of modulated and background signals, Fourier analysis etc.).

    If the reflectivity of an etalon structure is modified by mirror coatings on the interfaces (e.g. if one side is a highly reflecting mirror like in a typical diode or microdisk laser resonator), additional phase shifts can appear and have to be taken into account. In passive layer structures with absorption or active media with significantly high gain coefficients, the transfer functions have to be further modified.

3.2.2. Fabry-Pérot etalon of space-variant thickness at monochromatic illumination

    For non-uniform single layers (as in the case of thickness-modulated thin-film microlenses or microaxicons), a spatial dependence of the interference is obtained. For small angles, the formulas for Fabry-Pérot etalons can further be applied in good approximation. The spatial dependence of the transmitted intensity of a layer with Gaussian-shaped thickness distribution is demonstrated in Figs. 18 and 19. The calculation (with the same material parameters as used in Fig. 17) shows how the

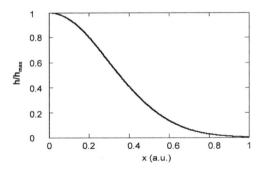

Fig. 18. Gaussian-shaped thickness profile of a layer structure used for the simulation in Fig. 19.

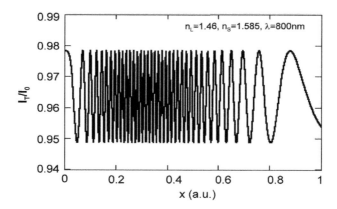

Fig. 19. Simulated intensity transmission of a layer with Gaussian thickness profile as in Fig. 18 ($n_L = 1.46$, $n_S = 1.585$, $\lambda = 800$ nm, maximum layer thickness 10 μm).

distance of the fringe maxima changes in correspondence to the thickness whereas the minimum and maximum transmission values are the same as before. One can conclude that a variation of the thickness profile function opens the possibility to spatially adapt the fringe density if this is necessary for specific applications.

3.2.3. Fabry-Pérot etalon of space-variant thickness at polychromatic illumination

In the case of polychromatic illumination, the summation over the E-field components has to be performed not only for all reflected or transmitted partial beams but also for all of their *spectral contributions*. In the numerical simulation, the fields have to be calculated for a sufficiently high number of discrete wavelengths. The square of the complete sum of the transmitted fields delivers the transmitted intensity [2001_10]. The numerical stability has to be verified by the convergence of the results for increasing numbers of wavelength slices. Figure 20 shows the field transmission for 4 different spectral components through the etalon of Gaussian thickness profile (see Fig. 18) illuminated by a polychromatic beam with a spectrum concerning to Fig. 21.

The envelope function of the spatio-spectral distribution was generated by adding half a period of a sinusoidal function (amplitude 1) to a constant value (1) over the distance of interest (x-interval see Fig. 18). The integrated intensity in Fig. 22 shows that the spectral interference of broadband signals in phase modulated microstructures can lead to significant inhomogeneities (spatial beating) which correspond to the spatially variable spectral profiles.

In particular, local contrast reduction or enhancement by destructive or constructive interference can be observed. On one side, such effects have to be taken into account as

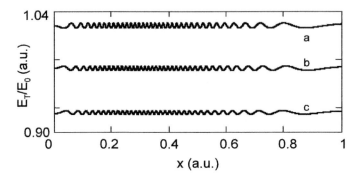

Fig. 20. Spectral interference of electrical field components for a Gaussian-shaped etalon (see Fig. 18) at polychromatic illumination. The curves correspond to the transmission at 3 different wavelengths within a given spectral distribution (see Fig. 21): (a) 770 nm, (b) 850 nm, (c) 900 nm. (Resulting beat structure see Fig. 22.)

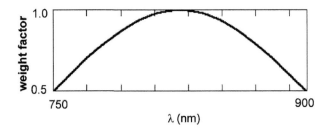

Fig. 21. Spectral distribution used for the simulation of spectral beating (see Fig. 22) at polychromatic illumination. The spectral range of 150 nm bandwidth was numerically sampled in steps of 10 nm. From the envelope, wavelength dependent weight factors for the field components were derived.

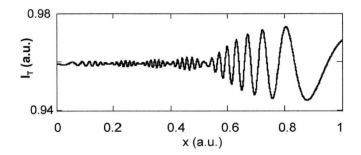

Fig. 22. Spectral distribution used for the simulation of spectral beating (see Fig. 21) at polychromatic illumination. The spectral range of 150 nm bandwidth was numerically sampled in steps of 10 nm (15 discrete wavelengths).

possible sources of errors in systems for beam shaping, measuring or data transmission. On the other side, the effects can be exploited to analyze or to shape field patterns. In the case of ultrashort wavepackets with broad spectral bandwidths and short coherence time, the problem of spectral interference has to be further generalized from the spatial to the spatio-temporal domain.

Polychromatic interference effects in thin films are of relevance for classical interferometric situations with white-light illumination as well. Recently, it was shown that the influence of multiple reflections on white-light interference microscopy has to be taken into account in the interpretation of fringe visibility distributions [2004_28].

3.2.4. Transmission of ultrashort pulses through plane-parallel etalon structures

The temporal transfer of ultrashort pulses through etalons depends on the pulse duration and the optical path length within the etalon. The necessary minimal condition for the generation of interference in any interferometer structure is the existence of a zone of significant overlap of the coherence volumina of two or more partial waves. The situation is schematically sketched in Fig. 23.

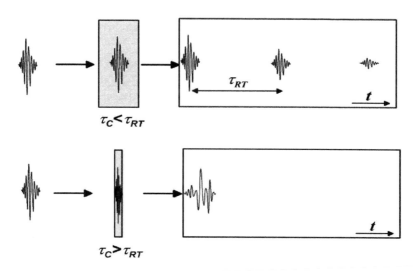

Fig. 23. Problem of the transmission of ultrashort light pulses through thin etalons (schematically). Depending on the ratio between coherence time τ_C and round trip time τ_{RT}, pulse trains of decaying intensity (from pulse to pulse, above) or single pulses of lengthened and deformed temporal profiles are produced (below). In the picture, the round trip time was changed by varying the thickness of the etalon. The boxes on the right side represent the temporal evolution after passing the etalons.

If we express the problem by the roundtrip time of the etalon

$$\tau_{RT} = \frac{L_{RT}}{c} = \frac{2hn_L}{c}$$

(52)

(c = velocity of light in vacuum) and by the coherence time τ_C of the pulse, the interference condition can be written as

$$\tau_C(\lambda_i) > \tau_{RT}(\lambda_i)$$

(53)

for all spectral contributions λ_i between minimum and maximum relevant wavelengths λ_{imin} and λ_{imax} in the spectrum. Without interference ($t_C \ll \tau_{RT}$), the pulse will be split into a train of temporally separated pulses of the repetition frequency:

$$\nu_{Train} = \frac{c}{2hn_L} \quad .$$

(54)

In the other extreme case ($t_C \gg \tau_{RT}$), the etalon acts as a spectrally selective interferometer and the pulse is stretched depending on the spectral interference (see [1992_5, pp. 51-53] and references cited there).

3.2.5. Quarter-wave layers and half-wave layers at monochromatic illumination

As it was already pointed out (see eqs. (34) and (35)), the phase of a monochromatic wave after passing a dielectric, non-absorbing layer at normal incidence depends on the optical thickness of the layer. For thin-film optical design, two special cases are of extraordinary importance [1995_26, p.7]. The first one corresponds to an optical thickness of a quarter of a wavelength (*quarter-wave layer*), the second one to an optical thickness of half a wavelength (*half-wave layer*).

For an optical thickness $n_L h = \lambda/4$, we obtain from eq. (34) a phase difference $\Delta\Phi_{QW}$ between the two beams reflected at the surfaces of a layer and the substrate-layer-interface (Fig. 24a) as well as between the transmitted beam and the beam reflected at the substrate-layer interface (Fig. 24b):

$$\Delta\phi_{QW} = 4 \cdot \frac{\lambda}{4} \cdot \frac{\pi}{\lambda} = \pi$$

(55)

The destructive interference at this phase difference is exploited to reduce unwanted reflection (single layer anti-reflection coating). As in all interference processes with monochromatic light, the minimum intensity is obtained in the case of equal amplitudes. For a given substrate material of the refractive index n_S, the layer index for balancing the amplitudes is:

$$n_L = \sqrt{n_S n_A} \tag{56}$$

with n_A as refractive index of the ambient medium in contact with the layer (e.g. air).

In the opposite case of $n_L h = \lambda/2$, a phase difference of:

$$\Delta\varphi_{QW} = 4 \cdot \frac{\lambda}{2} \cdot \frac{\pi}{\lambda} = 2\pi \tag{57}$$

leads to constructive interference so that the losses by reflection are finally added up again and the wave is not effected at all. Of course, this is only valid in the case of continuous waves (see the previous paragraph).

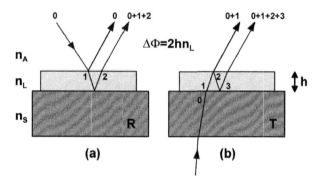

Fig. 24. Path difference of interfering partial beams for the design of quarter-wave and half-wave layers: (a) reflecting arrangement, (b) transmitting arrangement (schematically).

3.2.6. Admittance of absorbing layers at optical frequencies

In analogy to the theory of the conduction of electrical signals, the transfer of light waves through layers and layer systems can also be described by means of an *admittance Y*. It is the ratio between the magnetic and electric field amplitudes E and H of a harmonic wave propagating through a medium [1995_26, p. 2]:

$$Y = \frac{E}{H} \ . \tag{58}$$

At optical frequencies, the magnetic component can be neglected and the admittance of the layer Y_L can be approximated by the product of the refractive index n_L and the admittance of vacuum Y_0:

$$Y_L = n_L Y_0 \tag{59}$$

The value of Y_0 is 1/377 S (S is the SI-unit Siemens, 1 S = 1 A/V). For practical calculations, Y_0 is commonly normalized and measured in n_L free space units [FSU] [1995_26, p. 3]:

$$y = \frac{Y_L}{Y_0} = n_L \cdot [\text{FSU}] . \tag{60}$$

The admittance of an absorbing layer with the extinction coefficient

$$\kappa = \frac{\alpha_L \cdot \lambda}{4\pi} \tag{61}$$

(α_L - absorption coefficient of the layer material, λ - wavelength) can be written as

$$y = (n_L - i\kappa)[FSU] = (n_L - i\frac{\alpha_L \cdot \lambda}{4\pi})[FSU] . \tag{62}$$

Phase and absorption are related to each other via the Kramers-Kronig relation [2005_32]:

$$n_L(\omega) = 1 + \frac{c}{\pi} \cdot P \cdot \int_0^\infty \frac{\alpha(\Omega)}{\Omega^2 - \omega^2} d\Omega \tag{63}$$

(ω, Ω - angular frequencies, P - permittivity). In the last formula, the wavelength λ was replaced by the angular frequency:

$$\omega = 2\pi \frac{c}{\lambda} . \tag{64}$$

The Kramers-Kronig relation represents an expression for the dispersion properties of the layer material and enables to determine the complex refractive index by a simple absorption measurement. Refractive index data can be approximated by the Sellmeier formula [1984_2].

The intrinsic intensity absorption A_L within a layer of the thickness h_L and the absorption coefficient α_L (neglecting the interference effects, single pass) is given by the Lambert-Beer's law:

$$A_L = \frac{I}{I_0} e^{-\alpha_L h_L} . \tag{65}$$

On the basis of the admittance data for incident, reflected and/or transmitted waves (y_0, y_r, y_t), the *amplitude reflection coefficients r* and *amplitude transmission coefficients t* can be calculated again [1995_26, p. 3-4]:

$$r = \frac{E_r}{E_0} = \frac{y_0 - y_r}{y_0 + y_r} \tag{66}$$

$$t = \frac{E_r}{E_0} = \frac{2y_0}{y_0 + y_t} \quad . \tag{67}$$

For absorption-free incident media, the *reflectance R* and *transmittance T* can now be written as functions of the admittances as well:

$$R = r \cdot r^* = \left| \frac{y_0 - y_r}{y_0 + y_r} \right|^2 \tag{68}$$

and

$$T = \frac{\mathrm{Re}\, y_t}{y_0} t \cdot t^* = \frac{4y_0 \,\mathrm{Re}\, y_t}{|y_0 + y_t|} \tag{69}$$

(r^*, t^* denote the conjugate-complex values of r, t).

The admittance model enables to describe the light propagation through (even complicated) *systems of thin films* in a mathematically elegant way as it will briefly be shown in the next section.

3.3. Dielectric multilayer structures

3.3.1. The multilayer approach

The proper simulation of light interference in stacks of dielectric layers (schematically drawn in Fig. 25) is of enormous practical impact not only for the commercial mass production of high-quality coatings and filters but also for the design of new and sophisticated components for complex tasks of spatio-temporal beam shaping.

With multilayers, the tailoring of spectral reflectance and transmittance properties is much more flexible compared to the single layer design because of the additional degrees of freedom (number of layers, materials, composition, gradients of thickness or index etc.). With respect to the modeling and fabrication, multilayers are more demanding as well.

3.3.2. Method of characteristic matrices

One of the most popular simulation methods was introduced by Abelés [1950_1] and later applied to many types of multilayers (see, e.g., [1984_5]). The used model is based on the separation *of two partial waves* propagating in *opposite directions*, a negative-going wave and a positive-going one. The transfer from the rear interfaces of the layers to the forward interfaces is described in a compact manner by a so-called

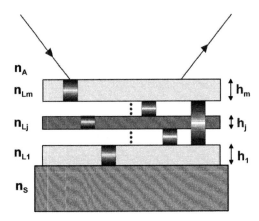

Fig. 25. Dielectric multilayer coatings for the tailoring of spectral reflectance and transmittance (schematically). The transfer functions result from the multiple interference between all different substructures (thicknesses h_j, refractive indices n_j).

characteristic matrix. The characteristic transfer matrix for the j-th layer at normal incidence at a wavelength λ is denoted [2003_26, p. 83] as

$$\mathbf{M}_j = \begin{bmatrix} \cos\phi_j & (i/n_{Lj})\sin\phi_j \\ i \cdot n_{Lj}\sin\phi_j & \cos\phi_j \end{bmatrix} \tag{70}$$

where Φ_j means the phase of the particular layer:

$$\phi_j = \frac{2\pi}{\lambda} n_{Lj} h_j \tag{71}$$

(n_{Lj} - refractive index of the j-th layer, h_j - thickness of the j-th layer, λ - wavelength in vacuum). For *multilayers*, the characteristic matrix is the product of the characteristic matrices of all contributing layers [1995_26, p. 4]:

$$\mathbf{M} = \begin{bmatrix} B \\ C \end{bmatrix} = \left\{ \prod_{j=1}^{q} [\mathbf{M}_j] \right\} \begin{bmatrix} 1 \\ y_s \end{bmatrix} \tag{72}$$

B and C stand for the total electric and magnetic field amplitudes. The ratio

$$Y = \frac{B}{C} \tag{73}$$

is the effective admittance of the coating which acts on the incident wave. With the help of the characteristic matrix method, *effective optical constants* of multilayers can be derived [2003_12]. Because of the equivalence of the structure of the characteristic matrices and single layer matrices, so-called equivalent single layers were introduced by Herpin [1947_1]. It was later shown that equivalent layers are closely related to the coupled-mode theory (which originally was used to describe waveguides, fiber gratings and distributed-feedback lasers) [2000_29]. In an alternative picture, the equivalent phase thickness is related to the Herpin model [1995_29].

Recently, significant progress in the method of characteristic matrices was obtained by reducing the representation to three basic matrices. On this basis it was possible to sort multilayers in three classes [2001_33]. By disclosing relationships to hyperbolic geometry, further useful mathematical tools became available [2000_30, 2002_40].

If we remember paragraph 3.2.5. and apply the matrix method to the transfer behavior of *quarter-wave and half-wave layers* in the monochromatic case, the problem finally takes the simple shape of the so-called *quarter-wave rule* [1995_26, p. 7]:

$$y_{QW} = \frac{y^2}{y_S} \tag{74}$$

where y is the admittance of the layer (see eq. (62)) and y_S the admittance of the substrate. The admittance of half-wave layers (after repeating the quarter-wave matrix operation) is

$$y_{HW} = y_S \ . \tag{75}$$

This means that in the result of the interference the wave "lost the memory" for any reflectance information.

Recently it was demonstrated that the optical admittance can also be used to compensate for thickness errors of nonquarterwave layers during the deposition [2006_23]. This enables to fabricate even complicated many-layer systems which are of essential interest in modern laser optics. Selected applications for the temporal and spatial control of laser radiation will be presented in the next paragraph.

3.3.3. Dispersion management by layer stacks of adapted spectral phases

With stacked layers, the cases of high and low reflectance can be realized as well. However, contrary to the monochromatic case, the enhanced number of degrees of freedom allows even for a broadband spectral management. One of the most challenging applications of multilayers is the dispersion control of ultrashort laser pulses. The study of the frequency dependence of the reflected phase [1978_3, 1990_8, 1993_23] resulted in the idea to control the group delay dispersion (GDD) via an *adapted penetration depth* of different spectral contributions in *non-resonant multilayer mirrors* [1994_27]. This is the basic principle of the so-called *chirped mirrors* (CM) which in many cases replaced classical compression systems with prism or grating pairs [1997_35, 2000_28]. The schematic of the design of a CM is depicted in Fig. 26.

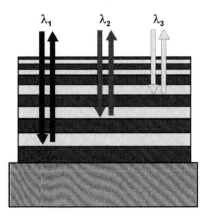

Fig. 26. Chirped mirror for the dispersion control in broadband ultrashort-pulse systems (schematically). The composition and geometry of a stack is adapted to the spectral bandwidth. The thickness of radially uniform partial layers varies with the depth in the stack.

The thickness distribution of partial layers over the layer number follows one (in the case of the simplest CM) or two (in the case of doubly chirped mirrors, DCM) characteristic envelope functions. A detailed discussion of the problem of dispersion control can be found, e.g., in [2003_26, pp. 393-421].

3.3.4. Diffraction management by layer stacks of spatially variable reflectance

Whereas the above described chirped mirrors take advantage of the variable thickness and composition of uniform layers to control laser pulses in the spectral and temporal domain, other degrees of freedom are accessible by varying the thickness of one or more layers in radial direction. The first sort of graded multilayer components are the *graded reflectance mirrors* (GRM), also referred to as variable reflectance mirrors (VRM) [1989_19, 1995_26, pp. 475-519]. In such multilayer structures, a *radial variation of the thickness* of one or more layers or the complete stack [1989_4, 1989_18, 1990_9, 1991_3, 1991_4, 1995_3] leads to a radial change of the reflectance for a given wavelength (schematically drawn in Fig. 27). GRM are typically used as apodizing output-coupling mirrors in unstable resonators for high power carbon dioxide lasers, excimer lasers [1996_24, 1999_6] or Nd:YAG-lasers. The principle of apodization is the suppression of side-lobes of diffraction patterns ("foots") by appropriate filter functions in the pupil plane [1964_2]. Typical reflectance profiles for the realization of large diameter single transverse mode operation are Gaussian or super-Gaussian distributions. For the design of GRM with a geometrically similar variation of the thickness of all the layers, a fruitful design approach was to match the first minimum of the typically strongly oscillating reflectance curves to the rim of the substrate [1989_4].

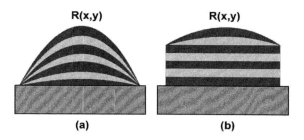

Fig. 27. Different types of graded reflectance mirrors (GRM) for the diffraction management in laser resonators by apodization (schematically). In GRM, the thickness of all layers (a) or a part of the layers (b) of the stack can be radially nonuniform [1995_26, pp. 479, 1996_24].

For applications in unstable laser resonators, different, contrary design criteria have to be fulfilled (e.g. a maximum filling of the medium on one hand but apodized output-coupling on the other hand). From a detailed analysis of the mode propagation it can be concluded that in many cases the optimum performance is obtained by applying *super-Gaussian reflectance profiles*. Here, the radial reflectance *R* can be written as

$$R(x) = R_{max} \cdot \exp\left[-2(x/w)^k\right] \tag{76}$$

(R_{max} - peak reflectance, k - super-Gaussian exponent, x - radial coordinate, w - waist radius) where the beam waist is defined by

$$R(w) = R_{max} \cdot \exp(-2) \ . \tag{77}$$

At high reflectance in the central region, local minima in the near field are caused. The optimum output profile ("maximum flat" condition) is a function of the magnification *M* of the resonator [1988_11]:

$$R_{max}(opt) = 1/M^k \ . \tag{78}$$

To optimize real laser systems with temporally and spatially variable gain and losses, the theoretical results for the mode propagation in empty resonators have to be further modified.

The design of *broadband GRM* is rather complicated. The combination of chirped and graded mirror designs, however, would be a worthwhile task because both dispersion as well as diffraction control could be integrated in a single multilayer component. The fabrication technologies for GRM played an important role in the history of thin-film microoptical components. Therefore, we will return to this topic and go more into the details in the next chapter.

3.4. Metal and metal-dielectric coatings

3.4.1. Single reflecting metal layers

In contrast to dielectric layers, metal layers show much *higher extinction coefficients* (corresponding to low penetration depths of the field) and *low refractive indices* (corresponding to high field reflectance). Therefore, relatively thin metal layers are widely used as highly reflecting mirrors. Particular materials have a broad spectral distribution of the reflectance in near infrared range. At Ti:sapphire laser wavelengths around 800 nm, e.g., Al and good metals like Au and Ag work well as reflector coatings. However, the losses within real metal layers are a limiting factor for high-power applications [1995_26, p. 21]. A detailed discussion of the problems of the polarization-dependent reflectance of metal surfaces can be found in [1957_1, pp. 509-534].

Recent investigations of the scattering of few-cycle pulses at thin metal layers show, that the carrier-envelope phase (CEP) difference of the wavepacket has to be taken into account for such kinds of processes [2004_29]. It has to be concluded that for extremely short pulses, the (typically rather poor) spatial homogeneity of the pulses plays an increasing role also from point of view of CEP.

3.4.2. Metal-dielectric layers as Gires-Tournois interferometer structures

Combinations of dielectric and metal layers can theoretically be used as antireflection coatings but suffer from losses as well. For practical applications, interference effects in such structures are more important. They either can be exploited to manage the spectral dispersion or have to be taken into account as unwanted effects in special types of hybrid microoptical components. In 1964, Gires and Tournois proposed the use of a multiple layer for dispersion control [1964_1]. The so-called *Gires-Tournoir interferometer* (GTI) was similar to a Fabry-Pérot etalon where one interface is replaced by a HR-mirror (so that the device works always in high reflectance mode) and the second one by a partially reflecting dielectric multilayer mirror or single layer. The distance can be fixed by a spacer. The reflectance R of the partially transmitting mirror determines the group delay (GD) of the GTI as follows:

$$\tau_{GD}(\omega) = \frac{(1-R)\tau_{RT}}{1+R-2\sqrt{R}\cos(\phi-\omega\cdot\tau_{RT})} \qquad (79)$$

(τ_{RT} - round-trip time of the cavity) [2003_26, p. 395]. For the maximum possible GD for each resonance frequency of the GTI one finds:

$$\tau_{GD\max} = \tau_{RT}\frac{1+\sqrt{R}}{1-\sqrt{R}} \ . \qquad (80)$$

Out of the resonance, the GD is decreasing so that negative group delay dispersion (GDD) can appear. The disadvantages of GTI might be overcome with more sophisticated structures like multiple-cavity GTI (resonant dispersive mirrors).

3.5. Problems and trends

The techniques of layer design were improved during about half a century. *Evolutionary algorithms* even enable to mimic biological optimization strategies to approach target parameters. Inhomogeneous coatings with continuous index variation were developed to overcome the limitations of classical design, e.g. at extremely short wavelengths [2006_24]. The implementation of absorbing layers can be exploited to design broadband AR-coating [1997_36]. Broadband multilayer designs were obtained for oblique angles of incidence [1994_38, 1997_37], in omni-directional operation [2001_34] or at different spectral ranges from visible to infrared [1997_38] and were even applied to the optimization of nanooptical probes [2003_29]. New design methods based on a successive (layer-by-layer) procedure [2002_41] or Fourier transform techniques with frequency filtering [1995_28] are investigated. Another important problem is the design of broadband wide-angle single layer [1985_4] or multilayer polarizing beam splitters [1996_25] which are necessary for broadband lasers, ultrafast spectroscopy, and interferometric devices like autocorrelators.

Nevertheless, the complexity of systems with many layers, different materials, gradients or filigran substructures continues to be a challenging problem and requires highly capable hard- and software for a reliable computing and a further confinement of models. Currently, compositions of nanolayers for *extremely short wavelengths* or chirped mirrors for *femtosecond laser* compressors have to be optimized with highest possible accuracy [2003_26, pp. 393-421]. For the dispersion control of octave spanning or even broader spectra of ultrashort-pulse lasers, new types of mirrors like TFI (tilted-front-interface chirped mirror) or BASIC (backside-coated mirror [2000_7]) were developed. The further optimization and experimental realization is currently within the focus of many activities. The *combination of dispersion and diffraction management* (spatially graded chirped mirrors) and the *combination with micro-and nanostructures or adaptive optical approaches* seem to be interesting fields for innovative research.

Another class of effects of particular relevance for high-power applications are *nonlinear* interactions in layer components [2004_16] which urgently have to be taken into account in system design. The proper implementation in simulation software has not been finished until now and thus remains a task for the future. Nonlinear changes of refractive indices and other channels of multiple photon interactions can be a source of distortions but also might offer specific solutions for next-generation spatio-temporal beam shaping problems.

The complex relationships between intrinsic structures of sculptured layers [1997_34, 2000_24, 2005_1] and their functionality is another promising topic which should be mentioned. The same has to be said about the study of biological layer structures, in particular with color and/or *polarization management* (e.g. anti-polarization coatings based on sculptured multilayers in butterfly wings [2001_32]).

Chapter 4

THIN-FILM MICROOPTICS

4.1. The concept of thin-film microoptics

In the previous sections, two basically different approaches for tailoring light fields in space and time were introduced: micro-optics and thin-film optics. *Microoptical* components interact with light typically by *reshaping the angular distribution* via refraction, reflection or diffraction at surface profiles or transversal index gradients. The resulting beam deflection is approximately localized in transversal direction like in the picture of "ray" optics [2001_13]. The structural information of microoptical beam shapers is encoded in spatial frequencies and transformed in wavefront curvature distributions in spatial domain (schematically in Fig. 28a).

Contrary to this, *thin-film interference structures* exploit the *internal interference* in single or multiple layer structures to control the spectral phase and thus reshaping reflectance or transmittance profiles (Fig. 28b). Here, the effects are well-localized in propagation direction (resonances in very thin layers). The structural information of optical paths to resonance lengths is transformed to filter characteristics in spectral and temporal domain.

The concept of *thin-film microoptics* takes advantage of both approaches. It *combines* the specific structural and/or functional features as well as advantages of fabrication and design procedures with each other. Furthermore, it enables to realize components and systems of significantly *extended functionality*, as it is schematically (and without claiming to be complete) sketched in Fig. 28c. With such structures, theoretically proposed generalized types of optical elements which have been referred to as *"refractive-diffractive-dispersive structured optical elements"* (Rafael Piestun, [1994_6; 2001_16]) can be physically approximated.

The control of phase and amplitude of light fields at extreme parameters, in particular the shaping and diagnostics of ultrafast, ultrabroadband, short-wavelength or near-field signals is the key to optical systems of higher resolution in space, time and spectrum. Therefore, a considerable impact of thin-film microoptics on future technologies has to be expected.

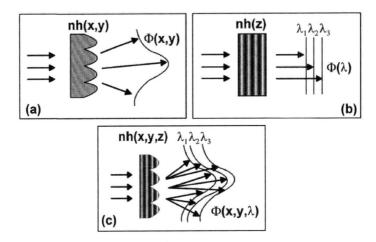

Fig. 28. Thin-film microoptical approach to a generalized shaping of the amplitude and phase (schematically). The functional features of conventional microoptics (a), where a spatially variable phase is typically used to transform wavefront and field amplitude by refraction, reflection or diffraction, are combined with those of multilayer interference components (b), where flat single or stacked substructures shape reflectance or transmittance profiles by selecting the spectral phase and amplitude with respect to their resonance behavior. The parameter *nh* (optical path length) is a function of transversal and/or axial coordinates.

4.2. Techniques for the fabrication of structured thin films: from macroscopic to microscopic scale

4.2.1. Subtractive and modifying techniques

There are two basic approaches to fabricate thin-film microoptical structures: (a) the initial deposition of a thin layer or layer system followed by a subtractive or modifying *structuring processes*, and (b) a *direct deposition or growth* of a layer with a spatially variable rate proportional to the final structure (additive process).

Examples for an type-(a)-process are the liquid-phase photolysis of sol-gel layers, the lithographic sculpturing of thin layers [1990_1, 1993_7] and layered waveguides [1971_2, 1974_3], the laser ablation of multilayers from front as well as rear side [1995_25] and the laser densification of optical films [1992_14]. Already in 1972, d'Auria et al. from Thomson-CSF reported on the lithographic fabrication of an array of 32 x 32 miniature lenses of 2 mm element diameter (focal length 13 mm, optical efficiency 35% at λ = 900 nm) [1972_1]. Later, the lithographic methods were applied to more complex DOE structures [1994_20, 1994_24, 1994_33, 1994_34].

Type-(b)-techniques are laser chemical vapor deposition (LCVD) [1987_7, 1990_2], laser physical vapor deposition (LPVD) [1992_16] and space-variant vacuum deposition through shadow masks. The quality of microlenses deposited with such procedures is typically not sufficient for applications in imaging systems (aberrations, absorption, scattering). Therefore, these methods were of only moderate interest up to now.

In the next paragraphs, the historical development of selected deposition techniques of thin-film microoptics and problems of mask shading will be highlighted. The main emphasis will be laid on vacuum deposition with *thick shadow masks in rotating systems*.

4.2.2. Uniformity and nonuniformity of deposited layers

Shadowing techniques for shaping distributions of physical matter as well as techniques to obtain defined profiles of light beams are well-known for a long time. In the case of vapor deposition, the angular distribution of a particle beam in a vacuum chamber significantly influences the thickness profile of the layer. For many practical applications there exist two typical, just complementary situations: (a) to deposit layers of highest possible *uniformity* (HR-mirrors, AR-coatings, spectral filters) or (b) to control the thickness parameters over an area to obtain a defined *gradient* (nonuniform) phase and amplitude filters [1965_2], *graded reflectance mirrors* (GRM) [1974_2, 1974_4, 1983_2, 1984_6], or graded spectral filters [1971_3]).

The use of point-like vaporizers in vacuum deposition systems leads to a certain nonuniformity of the layer thickness profiles because of the cos^x-dependence of the angular vapor beam profile [1952_1, 1972_2, 1987_5, pp. 76-77, 2002_37, pp. 489-522]. The *uniformity* can be improved by masks [2000_31] and moving the substrates. A monitoring of the deposition process is possible by measuring the transmission of the layer systems [1972_2, 1972_3, 1972_4, 1982_1, 1989_17, 1993_23]. By *rotating* the substrates, large areas could be coated with fairly homogeneous layers. (To get an idea of the technical challenges one should imagine that thin films have currently to be controlled within an accuracy of up to 10^{-4} over diameters of up to 10^8 times of their thickness!)

The inverse problem was triggered by new requests from the technical optics. In order to enhance the resolution of microscopes or astronomical instruments, to spatially select spectral lines, or to improve the performance of high-power lasers (mode selectivity, mode volume, frequency and pointing stability etc.), diffraction effects had to be minimized (apodization [1962_2, 1964_2]). For this purpose, one has to suppress any hard edges in the system. This can be realized with components of adapted *spatially nonuniform* amplitude or phase characteristics. The question was: How to generate the necessary profiles?

4.2.3. Separately rotating masks for structured light exposure and vapor deposition

The first attempts to generate radially symmetric *apodizing elements* where not based on thin-film techniques but on *photographic procedures* which belong to the modifying techniques. It was found that apodizers for low-power applications can simply be generated by exposing photographic layers. The trick was to introduce *rotating slits* of well-designed shape to illuminate the specimen with a *radially-variable exposure time* [1962_1]. Because of the necessary distance between rotating masks and substrates and unavoidable limitations of the accuracy, in the central part appeared "bad" areas of up to 1 mm diameter so that the mask-shadowing method works well for larger diameters (cm-range) but is less applicable for microscopic-size elements.

The rotating mask approach was later *adapted to the deposition techniques*. The setup is schematically shown in Fig. 29. Macroscopic-size GRM (see Chapter 3) of mostly Gaussian or super-Gaussian reflectance profile were obtained in this way [1964_3, 1975_2, 1981_2, 1980_2, 1989_19, 1990_10, 1995_26, pp. 475-519]. Here, a mask with well-shaped slit-apertures rotates in close proximity to the substrate as well. In contrast to the photographic process where in principle graytone masks can be applied, shading systems for vapor beams have to have real openings so that the material can pass through.

The technique of rotating masks is absolutely suitable to accurately shape the thickness profiles of large-area components. However, as in the case of photographic exposure, its capability for a miniaturization is rather limited because of the problem

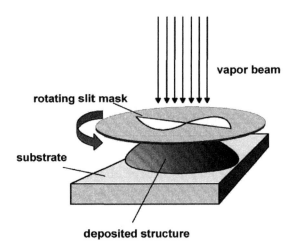

Fig. 29. Deposition technique with single rotating mask containing a slit aperture (schematically). The shape of the slit is optimized to generate a certain thickness profile.

of reduced accuracy near the rotational axis. Furthermore, a simultaneous production of multiple elements (e.g. of arrays) is very difficult because of the relative rotation between mask and substrate (the application of high-precision micro-mechanical masks might lead to an improvement but seems to be expensive).

For smaller structures (diameters < 1 mm) and large numbers of elements (extended arrays), techniques with fixed masks are basically more adequate. Therefore, they are very attractive for microoptical and nanooptical structuring.

4.2.4. Fixed thick shadow masks and extended sputtering sources

By *sputtering* with thick (3D) shadow masks at fixed positions relative to the substrate, single Luneburg lenses [1981_1] for integrated optics were fabricated by the group of Yao [1978_1, 1978_2, 1979_1]. The scheme of the arrangement is shown in Fig. 30. Here, the sputter *source* was *extended* (contrary to the approximately point-like sources in vacuum deposition systems with electron beam heating). The deposition process was simulated numerically by taking into account the spatial extension of the source and the depth profile of the mask. By integrating over all contributions of the vapor beam at each point on the surface to be coated, the resulting lens profiles were determined. The shape of the mask was optimized to generate the wished distribution.

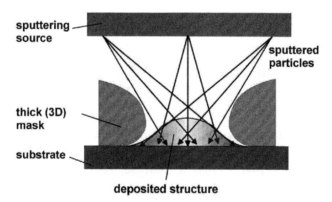

Fig. 30. Deposition technique with a thick (depth-shaped) mask and an extended sputtering source (schematically).

4.2.5. Fixed thin circular shadow masks rotating with the substrate

Thin shadow masks fixed at a certain distance to the *rotating substrate* were first used to deposit soft diaphragms for the apodization in laser systems [1974_4, 1976_1]. A particular type of amplitude filters were such with inverse Gaussian profiles [1985_5]. One of the problems to be solved was to overcome hot spots of high-power lasers by mode cleaning (e.g. in systems for laser fusion). The technique was later confined and applied to the deposition of mirrors of Gaussian and super-Gaussian reflectance of the type shown in Fig. 27b (Chapter 3) where only one or several particular layers of a multilayer are thickness modulated [1988_14, 1988_10, 1988_11, 1988_12, 1998_5]. A serious problem was to suppress the residual side-lobes of oscillating profiles [1988_14, 1988_13]. The disadvantage of the method is that the apparatus has to be opened between depositing uniform and nonuniform layers. Otherwise, a mask has to be steered by a remote control what makes the procedure more laborious as well.

4.2.6. Thick shadow masks fixed at the substrates in a rotating system with point source

An alternative design solution for the deposition of GRM was to proportionally shape a *complete layer system* (Fig. 27b in Chapter 3) with *shadow masks fixed at the substrates*. An important advantage is that the vacuum chamber can be closed over the complete process. Symmetric and homogenized distributions can be obtained by

(a) rotation of the substrate around its own axis,
(b) bound rotation on a circular orbit,
(c) planetary rotation around a central axis.

The cost-effective fabrication of GRM for CO_2-lasers ($\lambda = 10.6$ µm), Nd:YAG-lasers ($\lambda = 1.064$ µm) with all-layer-modulation in systems with *planetary rotation* was demonstrated [1990_9, 1991_3, 1991_4, 1991_24, 1995_3, P1]. The arrangement is drawn in Fig. 31. Typically, bored metal plates or tubes of some mm height were applied as shadow masks [1990_9]. The 3D masks can also be replaced by stacks of thin masks. The optical thickness of the films was monitored during the evaporation process with a single-beam spectral photometer. This way, apodizing mirrors and diaphragms with diameters between 1 and 50 mm were produced with circular-symmetric, but also square- and stripe-shaped structures. The latter would be strictly impossible with techniques based on a slit rotating relatively to a substrate.

The design approach was found to be less complicated compared to multilayers with a spatial variation of only a part of the layers. The problem of oscillating

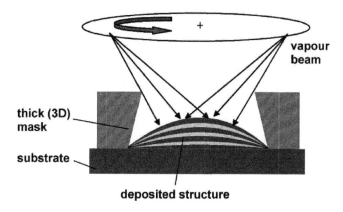

vapour
beam

thick (3D)
mask

substrate

deposited structure

Fig. 31. Deposition technique with a thick (depth-shaped) shadow mask and a point-like vaporizer in a rotating system where the mask is fixed at a given distance to the substrate (schematically).

reflectance, however, can be more complex. Furthermore, significant phase differences are generated. These *lens-like action* has to be taken into account in resonator design (but also can be exploited for beam shaping with *amplitude-phase-elements*) [1991_4].

The degrees of freedom of *thick masks* were exploited in a similar way as in the case of the sputtering setup. However, a *point-like source* is used in this case instead of a transversally extended one.

4.2.7. Deposition of arrays of microoptical components with miniaturized shadow masks

The progress in structured deposition technologies was considerably stimulated by the development of new types of lasers:

- Compact short-pulse solid-state laser resonators required a *miniaturization* of GRM.
- For the coherent coupling of laser arrays or separated gain sections of a laser, matrix arrangements of miniature mirrors were of interest [1995_31, 1994_2, 1994_3].

In test depositions through small apertures, single sub-millimeter size dielectric multilayer GRM and apodizing reflecting apertures were obtained. By depositing multilayer structures through hole array masks with periods between 200 and 800 µm, multiple GRM were fabricated as well. These structures were referred to as graded reflectance micro-mirror arrays (GRMMA) [1993_4, 1993_5] and represent a first real physical approximation of generalized structured optical elements (see Fig. 28c). GRMMA were used to built up Talbot resonators [1994_2, 1994_3, 1995_3] which exploit the Talbot effect for mode selection and coupling (see paragraph 2.9.4.). GRMMA are hybrid microoptical components with reflective as well as refractive properties (lens-like action) so that they can simultaneously enable for output coupling and focusing. The next logical step was to replace the deposited multilayers by single dielectric layers to obtain arrays of purely refractive microoptical components [1995_2, 1995_31]. The deposition setup is schematically shown in Fig. 32.

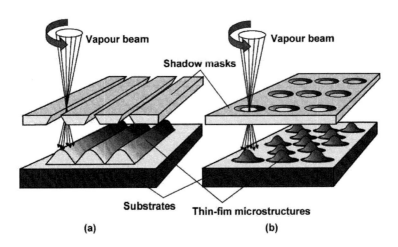

Fig. 32. Deposition of arrays of microoptical elements with multiple miniaturized shadow masks: Fabrication scheme of (a) cylindrical elements by shading with linear slit-array masks, (b) 2D arrays of circular elements with hole masks. The technique enables for depositing single layers as well as multilayer structures for phase or phase-amplitude beam shaping.

Arrays of slits and holes were used to fabricate arrays of cylindrical, circular or elliptical shape with pitches between 20 and 1000 µm in linear, orthogonal or hexagonal arrangements. Beside micro-mirrors and microlenses, also waveguides can be deposited. With shading techniques, uniform and nonuniform arrays (i.e. arrays with elements of different shape, size or thickness) can be obtained. Additional macroscopic thick masks (schematically in Fig. 33) allow to generate envelope thickness profiles [1997_5] which can be used to realize more sophisticated beam shaping functionality (e.g. spatially variable focus length or tailoring of the mode volume of supermodes in self-imaging resonators). Further degrees of freedom for a variation of local or global profiles arise from the distribution of diameters, shapes and arrangement of holes or slits in the array masks, and sequential deposition with spatial transformations of the system geometry.

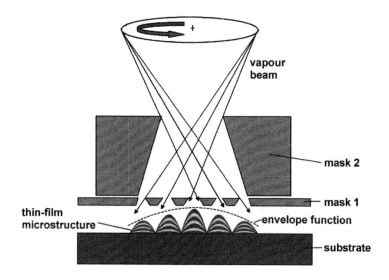

vapour
beam

mask 2

mask 1

thin-film
microstructure

envelope function

substrate

Fig. 33. Fabrication of thin-film microoptical structures with envelope thickness functions by additional macroscopic shadow masks (schematically).

4.2.8. Simulation of the deposition through thick shadow masks fixed on a substrate in a system with planetary rotation

For a simulation and optimization of the shading conditions, the deposition geometry including the rotation of substrate and mask, angular distribution of the vapor beam, distances between rotation axes and planes, number of rotations etc. have to be taken into account [1990_9]. The angular deposition of the vapor beam has to be determined experimentally by test depositions and can slowly vary, in particular for exhausted vaporizer material (self-shading). During a planetary rotation (characterized by two rotation axes), each substrate moves on a so-called *hypotrochoidal* curve [1977_1, 1995_30, p. 319] (Fig. 34). In the picture, the small circle symbolizes the substrate and the large circle the outer envelope of the orbit. The highest layer symmetry is obtained for a large number of rotations and closed cycles. Vice versa, asymmetric layers can be generated by breaking the system symmetry. If the substrate is placed at a central position in the rotating system, the planetary rotation degenerates to a simple rotation around a single axis. In the coordinate system presented in Fig. 34, this case corresponds to constant coordinates of $(x,y) = (0,0)$ for the center of the substrate (the outer circle diameter is then reduced down to the diameter of the substrate).

For reasons of simplicity, the simulation shall be explained here for the case of a symmetric, cylindrical shadow mask fixed at a substrate. We assume a rotation of substrate and mask at a central position in a first plane and a vaporizer at a certain

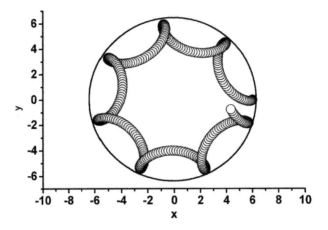

Fig. 34. Orbit of a substrate rotating in a deposition system with planetary rotation. The center of the substrate (small circle) moves on a hypotrochoidal curve.

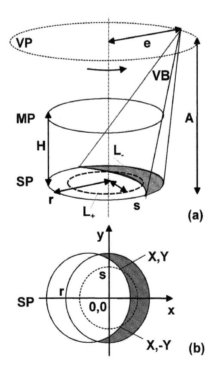

Fig. 35. Geometry of a shadow zone for a rotating cylindrical mask (*H* - height of the cylinder, *VB* - vapor beam, *SP* - substrate plane, *MP* - plane of the upper edge of the cylinder, schematically), (a) side view, (b) top view on the substrate plane.

distance to the rotation axis in a second plane [1990_9], The corresponding geometry of the crescent-shaped shadow zone is schematically drawn in Fig. 35. Here, each point on the substrate moves on a circle. The ratio of the paths within the shadow zone L_+ and outside of the shadow zone L_- is proportional to the effective deposition time and thus determines the local profile thickness h. The distance dx between the inner wall of the cylindrical aperture and the rim of the shaded zone is defined by the height of the cylinder H and the average angle of incidence $\langle \beta \rangle$ of the particle beam and can be calculated by

$$dx = H \cdot \tan\langle \beta \rangle \tag{81}$$

where the average angle is a function of the main radius of the substrate around the central rotational axis $\langle r \rangle$ and the distance between vaporizer and central axis e:

$$\langle \beta \rangle = \frac{1}{2} \cdot \left[\arctan\left(\frac{\langle r \rangle - e}{A} \right) + \arctan\left(\frac{\langle r \rangle + e}{A} \right) \right] . \tag{82}$$

The sums of all complementary paths $L_+(s)$ and $L_-(s)$ at radial coordinates s have to be:

$$L(s) = L_+(s) + L_-(s) = 2 \cdot \pi \cdot s . \tag{83}$$

As it can easily be shown, the arc length outside the shadow zone is

$$\frac{L_+}{2\pi} = \frac{1}{\pi} \cdot \arctan\left[\left(\sqrt{s^2 - X^2} \right) / X \right] \tag{84}$$

where X is the x-coordinate of the intersection points between circular path and the rim of the shadow zone:

$$X = \frac{r^2 - dx^2 - s^2}{2dx} . \tag{85}$$

The relative thickness, i.e. the ratio of the local thickness $h(s)$ after finishing the deposition to the thickness without any shading h_0, is directly proportional to the expression in eq. (84):

$$\frac{h}{h_0} = \frac{L_+}{2\pi} \tag{86}$$

so that we finally can calculate the local thickness as

$$h = \frac{h_0}{\pi} \cdot \arctan\left[\left(\sqrt{s^2 - X^2} \right) / X \right] . \tag{87}$$

The curves in Fig. 36 show corresponding thickness profiles for a particular typical set of deposition parameters. By changing the parameters of masks and deposition conditions, the resulting thickness profiles can be varied over a certain range. For applications in imaging microoptical systems and axial laser beam shaping, structures of *parabolic* [1999_2], *Gaussian* [2000_4] and *anti-Gaussian thickness profiles* [2004_11] and *arrays of high fill factor* are of particular interest. Therefore, considerable effort has been spent to approximate such target functions (see next Chapters). To enhance the design flexibility for the shape functions, additional tricks like sequential insertion of different masks, substrate translation, dynamic masks, overlap of rims of structures or phase shaping by multilayer interference can be exploited.

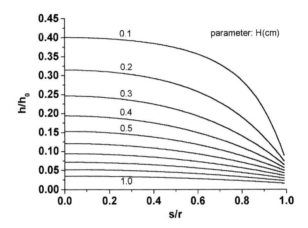

Fig. 36. Simulated thickness distributions for a deposition with a cylindrical mask at varying mask depth H (inner diameter of the mask $2r = 1$ cm, parameters of the vacuum deposition system: $e = 14.15$ cm, $A = 33.5$ cm, $<r> = 10.1$ cm).

By comparing the results of numerical simulations with experiments, the validity for patterns on the 100-1000 µm scale was verified. At small mask dimensions, errors of the mask (shape, distance) are more important but inhomogeneities of the vapor beam can be neglected. With etched high-precision masks, the production of structures in sub-50-µm diameter range is realistic.

Recently, the fabrication of *nanostructures* with shadow mask deposition techniques [2004_38, 2005_30, 2005_31] was demonstrated. The deposition of regular nanopatterns by arrays of shading nanospheres is also well known in lithographic structuring [2004_27]. Reliable data on lower spatial limits of the deposition, however, are not known up to now.

4.3. Specific properties of thin-film microoptical components

Compared to other methods, thin-film deposition with shadow masks has a number of specific advantages with respect to the fabrication procedure as well as structural and functional characteristics of the generated components. On the other hand, thin-film microoptical design is associated with several specific problems and limitations that have to be taken into account for the design. Both sides will be examined in this section.

4.3.1. Specific advantages of thin-film deposition technique with shadow masks

For good reasons, the deposition through shadow masks can be regarded as an enabling microoptics technology [1997_5, 2000_1, 2004_5, 2005_5, 2005_6]:

- Array masks enable for the *simultaneous deposition of a large number of microoptical elements*. The structured coating of substrates of 10 cm diameter with $> 10^4$ elements was demonstrated. Typical deposition systems can be used for a simultaneous coating of a certain number of substrates. Thus, a very *cost-effective production* is possible.
- Another advantage is the option to better exploit *spectral degrees of freedom* by including more than one material (multilayer design) as already discussed.
- In many cases, the fabrication in a *single process step* is possible.
- *Continuous relief structures* of high surface quality (*very low roughness*) can be generated without post-processing steps.
- Extremely flat structures with *low numerical apertures* can be realized without etching or reflow.
- The deposition on thin and mechanically *flexible substrates* (fibers, thin polymer foils) is easy.
- Compact microoptical systems can be realized by the *direct integration* of thin-film structures (e.g. in planar optical setups, sensors, filters, solar cells or on the face of laser diodes).
- Structures deposited on surfaces can either be used directly or *transferred into substrates*. Thus, an easy replication is also possible.

4.3.2. Specific advantages of thin-film microoptical design

The thin film design takes advantage from the specific properties of optical layers and layer systems which were already discussed in extenso. The most important peculiarities which make thin-film microoptics so attractive for laser beam shaping (in particular at really extreme parameters [2005_6]) are:

- The complete spectrum of basic functionalities of macroscopic thin-film optics can be transferred to the microscopic scale (HR, AR, partial reflection, dichroic or multichroic mirrors, spectral filters, apodizing mirrors and diaphragms, etc.) and combined with new functionality of microstructures and arrays.

- Very thin dielectric layers or purely reflective structures enable for systems with *low spectral dispersion* and therefore for undisturbed *broadband* temporal and spectral transfer of optical signals.

- By *low-absorption* or *reflective* microoptics, the risk of a damage of the components at high power densities can be minimized. Thermal expansion coefficients, refractive indices and spectral transmission can be well adapted by using identical materials for layer and substrate.

- *Spectral and spatio-temporal selectivity* can be obtained by dielectric multilayers, metal-dielectric composite systems or by exploiting the etalon properties of single layers on substrates of significantly different refractive index (see Chapter 3). Multilayer structures can be adapted to spatially variable parameters of the light wave to be shaped (e.g. frequency spectrum, angular distribution).

- Extremely *small surface tilt angles* and large radii of curvature enable to work under near-paraxial conditions. This feature is of increasing interest for the shaping of nondiffracting beams of large axial extension ("needle beams, see Chapters 6 and 7) and the manipulation of ultrashort-pulse wavepackets [2004_4] with minimum geometrical dispersion. Furthermore, phase masks with *small phase steps* and correction layers can be generated. Theoretically, the thickness can approximate zero. In practice, the minimum is limited by the loss in symmetry for low numbers of cycles and unclosed cycles. Other limitations arise from the obtainable substrate quality.

- Microoptics for *extreme spectral ranges* like VUV or IR can be fabricated either by transferring dielectric layers (e.g. fused silica) in materials of appropriate transmission characteristics (e.g. CaF_2), or by directly depositing layers on substrates of identical material (e.g. MgF_2 on MgF_2, ZnSe on ZnSe).

- *Waveguiding functionality* can be realized. (For example, the deposition through shadow masks can be used to deposit parts of waveguides consisting of dielectric layers and polymers.)

- *Matrix arrangements* of elements (e.g. microlens arrays or micro-mirror arrays) consisting of identical or variable elements can be adapted to a wide range of tasks like multichannel processing, coupling, beam homogenization or beam array generation.

4.3.3. Thin-film microoptics on substrates: Specific properties and design constraints

Because it is rather complicated to fabricate or to separate free-standing thin films, the typical component consists of the two parts:

- substrate (e.g. plane-parallel plate),
- layer (e.g. microlens).

From this arrangement result most of the specific properties of (and design constraints for) thin-film microoptical systems.

(a) Refractive index difference between transparent layer and transparent substrate

A non-zero index difference can be the reason for *parasitic reflections*, interference (see Chapter 3), and waveguiding so that unwanted Newton's rings or cross-talk are observed (schematically in Fig. 37).

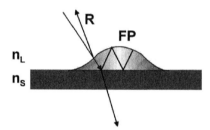

Fig. 37. Parasitic reflection (R) and internal Fabry-Pérot etalon effects (FP) in a thin-film microlens caused by differences between the refractive indices of layer (n_L) and substrate (n_S).

To reduce such influences, identical or minimal different indices can be used (e.g. fused silica on silica). For certain applications or measurements, even small internal reflections disturb the interpretation of results. For example, serious errors in fringe processing of interferometric profilometry at multilayer systems can be caused, e.g. parasitic oscillations, false phase steps and edge slopes, or apparent local thickness maxima. A possible loophole is the deposition of an additional highly reflecting, easy-to-eliminate auxiliary layer (e.g. a thin Au-coating) before executing the profile measurements. Another solution is the integration of AR-coatings in the design. If auxiliary coatings can not be applied, multiple wavelength measurements with a separation of relevant spatial frequencies by Fourier filtering can help.

At coherent illumination of transparent structures like in the case of coherent image processing or laser materials processing, such interference effects can be crucial because of propagating fringe structures in the near field. Parasitic interference from the substrate itself can even be observed with white-light interferometers if the thickness of the substrate is smaller than the coherence length.

(b) Layer adhesion on the substrate material

Depending on the substrate material and quality (cleanness, surface roughness) as well as on the layer properties (material, thickness, structure), a destruction or delamination of the layers can be caused by internal strain, thermal stress or bending load. Improvements can be obtained by *adhesion layers* (schematically in Fig. 38). Another way out is the transfer into the substrate where one ends up with a monolithic and robust structure.

Fig. 38. Improvement of the layer adhesion on the substrate by an adhesion layer (A) between substrate (S) and layer structure (L) (schematically).

(c) Bending of thin substrates

The bending of structures on flexible substrates (polymer foils, side-coated fibers) causes not only adhesion problems but also optical effects (aberrations, changes of focal lengths [1996_2], schematically in Fig. 39a, b).

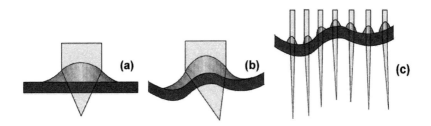

Fig. 39. Problems of substrate bending: (a) structure of large diameter on a plane substrate, (b) structure of large diameter on a bended substrate, (c) array of microstructures on a bended substrate (schematically).

Microlenses of long focal length can be more robust against tilt (see above discussed scaling laws for microoptics) so that they behave different (Fig. 39b) when bent. In particular at low fill factors, uncoated areas between the elements can well compensate for the mechanical forces. In general case, however, it is necessary to limit the radii of substrate curvature. (Vice versa, changes of the imaging properties can also be exploited for adaptive optical applications as it was demonstrated with deposition structured fiber-lenses for the smile correction at laser diode arrays [1995_32]).

(d) Substrate quality

For high-power laser applications, the *roughness, waviness and homogeneity* of thickness, refractive index and absorption of the substrate have to be well controlled to minimize distortions and scattering (Fig. 40). In particular at short wavelengths, scattering losses can cause significant cross-talk and losses (in UV 20% and more). At low layer thickness, the influence of phase errors of the substrate can not be neglected.

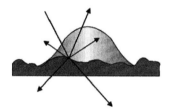

Fig. 40. Influence of roughness and waviness on the microlens performance

(e) Influence of the substrate thickness on the imaging conditions

In contrast to conventional optics and depending on the orientation of the microlens vertex (in propagation or opposite direction), the substrate also influences the imaging. Lenses and substrates have to be regarded to be two components of a common optical system. The corresponding refractive indices, thicknesses and radii of curvature of both parts define the focal length. Layer microlenses on a single side of a substrate have only a single curved face. For a plano-convex thin lens, the approximation in eq. (19) (Chapter 2) for the focal length f of a thin lens can easily be transformed to the well-known simple relation:

$$f = \frac{\rho}{n_L - 1} \tag{88}$$

(n_L - refractive index, ρ - radius of curvature, compare to [1974_5, p. 47]). The radius of curvature of a spherical calotte can be calculated from the thickness by the relation previously presented in eq. (14). If we add a substrate of a refractive index n_S in direct contact to the lens now, we have to analyze two different cases for the layer orientations (see Fig. 41).

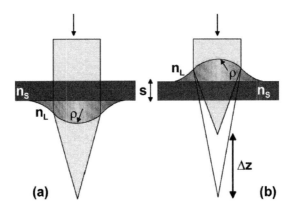

Fig. 41. Focus depending on the orientation of a plano-convex thin-film microlens on a substrate: (a) orientation in propagation direction, (b) opposite orientation.

(i) The first case corresponds to the *orientation in propagation direction* (Fig. 41a):

Here, smaller spherical aberrations appear [1974_5, pp. 227-228]. The focus position is not influenced by the thickness of the substrate and eq. (88) can be applied in good approximation.

(ii) In the second case, the substrate orientation is rotated by 180° (Fig. 41b):

Here the convex surface looks to the incident beam. The spherical aberrations are larger. At identical refractive indices of substrate and layer $n_L = n_S = n$ (optimum case), the expression for the focal length has to be modified to

$$f' = \frac{\rho}{n-1} - \frac{s}{n} = f - \Delta z \qquad (89)$$

(s - thickness of the substrate) and the focal length is effectively reduced by a distance of $\Delta z = s/n$. In the more general case of *different refractive indices of substrate and layer structure*, the direction of the shift of the focal spot depends on the ratio between both. Here, the following relation has to be applied [2000_1]:

$$f = n_S \left[\frac{\rho}{n_L - 1} - s \right]. \qquad (90)$$

For the expression within the bracket, one can introduce the constant

$$C = \left[\frac{\rho}{n_L - 1} - s \right] \tag{91}$$

and write

$$f = n_S C . \tag{92}$$

From the linear proportionality of the focal length to the refractive index of the substrate there follows an enhancement of the focal length for $n_S > n_L$ and a reduction of the focal length for $n_S < n_L$ compared to the case without substrate. The relative shift is

$$\frac{\Delta f}{f(n_L)} = \frac{(n_S - n_L)}{n_L} \tag{93}$$

and the absolute shift (difference of focal lengths with and without substrate) amounts

$$\Delta f = n_S \left[\frac{\rho}{n_L - 1} - s \right] - \frac{\rho}{n_L - 1} = \rho \frac{n_S - 1}{n_L - 1} - n_S \cdot s . \tag{94}$$

It has to be mentioned that also in the case of using transparent spacers (e.g. optical glue) or inserting glass plates in the optical path, a corresponding focal shift in axial direction has to be taken into account for microoptical system design. The so-called image shift b by a plane parallel plate of the thickness s and the refractive index n_S is [1997_5, p. 48]

$$b = \frac{n_S - 1}{n_S} \cdot s . \tag{95}$$

It is evident that in this case the distance of the focal plane from the lens is reduced by the distance b. Furthermore, for applications (e.g. large NA lenslet arrays in front of the pixels of an image sensor) it can be important that within the medium (e.g. glue) the angular distribution is changed (e.g. the minimum distance for a complete illumination of an active area is changed, the position of a black-matrix has to be modified, or multilayer coatings have to be adapted).

(f) Optical properties of the substrate material

The optical properties of the substrate (absorption, dispersion, polarization, birefringence etc.) are not less important, in particular for the shaping of high-power and ultrashort-pulse laser beams. For broadband applications, a reflective setup with multilayer or metal structures of large spectral bandwidth can be a solution. However, in many cases one has to pay the price of shadowing effects and off-axis design. Wavelength dependent transmission in visible range can be observed for certain polymers (e.g. a yellow color for polyimide). Because of their molecular structure, polymers tend to polarization effects and birefringence. Crystalline materials like CaF_2 (which is very important for short-wavelength microoptics) also are typically birefringent.

4.3.4. Contact angle and maximum angle of incidence of thin-film microlenses

The *contact angle* of thin-film microoptical elements to the surface of a plane substrate can be very small (e.g. in the case of nearly Gaussian-shaped elements) but also extremely large (as for high-NA spherical lenses). If a plane wave arrives in normal incidence to the substrate, the local incident angle depends on the surface tilt at the point of interest. The angular distribution is important for the design of multilayer optics, in particular if extremely high accuracy is required. For convex lenses, the minimum angle of incidence β_{inc} (related to the surface, not to the optical axis!) is observed at the rim which is the steepest part. For a hemi-spherical lens, the minimum and maximum incident angles are $0°$ and $90°$, respectively. The angle of incidence on the layer can be calculated from the contact angle β_{cont} by

$$\beta_{inc} = 90° - \beta_{cont} \ . \tag{96}$$

It is instructive to look at the likewise important case of a *parabolic thickness distribution*. From simple geometrical considerations, the contact angle of an ideal parabolic microlens as a function the ratio between the maximum thickness h_{max} and the microlens diameter d can be found to be [1998_17, p. 23]

$$\beta_{cont} = \arctan\left(\frac{4h_{max}}{d}\right) . \tag{97}$$

The dependence of β_{cont} on the relevant parameter d/h_{max} is shown in the logarithmic plot in Fig. 42. It can be recognized that large contact angles (and thus considerably small angles of incidence on the layer) appear which are all but negligible in multilayer design. If the illuminating light beam is divergent, the angular distribution of both the beam and the layer surface have to be regarded.

4.4. Types of thin-film microoptical components

The most important types of components which have been fabricated with shadow mask deposition methods are compared in the "family tree" of thin-film microoptics with respect to their optical functionality (Fig. 43).

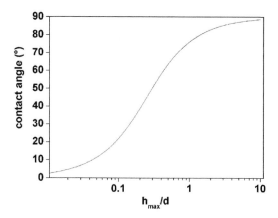

Fig. 42. Calculated contact angle of a parabolic microstructure (calotte) as a function of the ratio between the maximum thickness h_{max} and the microlens diameter d (logarithmic plot).

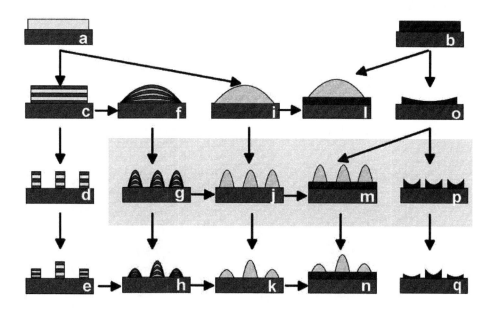

Fig. 43. Family tree of thin-film microoptics: (a, b) unstructured single dielectric and metal layer, (c) unstructured multilayer, (d) patterned multilayer (hard edges), (e) the same with envelope thickness function, (f) GRM, (g) GRMMA, (h) GRMMA with envelope, (i) dielectric layer microlens, (j) array of thin-film microlenses, (k) the same with envelope, (l) dielectric microlens on metal layer, (m) thin-film microlens array on metallic layer, (n) the same with envelope, (o) concave metal layer mirror, (p) array of metal layer micro-mirrors, (q) the same with envelope. The elements in the gray box represent the basic functionalities: refractive (j), reflective (p) and hybrid refractive-reflective (g, m).

In the schematic, the basic refractive, reflective and hybrid functionalities are addressed which directly correspond to specific properties of structured thin films. Other types of components like hybrid refractive-diffractive elements and gradient index layer structures are not included in the overview.

Beside the layer design, thin-film microoptics can also be classified by other criteria like their symmetry, thickness profiles, shape, array arrangements, uniformity or fill factor. The *symmetry* can either be broken with respect to the thickness profile (asymmetry in one direction, e.g. prism or saw tooth profile, or two directions, e.g. anamorphotic and cylindrical structures). The thickness profiles can be divided in *spherical and nonspherical profiles*. Nonspherical profiles are of essential importance for *axial beam shaping*, in particular the shaping of multiple focal spots, oscillating focal zones along the optical axis or so-called nondiffracting (Bessel-like) beams. These types of beams will be explained further below.

4.5. Fabrication of thin-film microoptics with shadow masks

4.5.1. Vacuum deposition with planetary rotation and shadow masks

Fig. 44 shows a typical commercially available vacuum deposition apparatus (Hochvakuum Dresden, model B55) with electron beam vaporizer and a home-made planetary gear which has been used by the group of the author in close collaboration with the team of Dr. Schäfer (Quarterwave company, now part of Berliner Glas) to

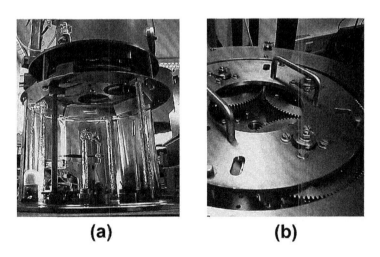

(a)   **(b)**

Fig. 44. Vacuum deposition apparatus type B55 (a) and planetary gear for rotating substrate holders (b) for the fabrication of thin-film microoptical components with shadow masks.

fabricate thin-film microstructures [1998_17, p. 28]. With this device, rotational speeds (with respect to the main rotational axis) of 10-50 cycles/min were driven (typically 30 rotations/min). The layer thickness was in-situ monitored by the interference of test layers at a central position in the system. The angular distribution function f(φ) of a vapor beam (fused silica) was determined by coating test layers on a set of 10 substrates at different equidistant positions (difference 3 cm) in one and the same plane without rotation and additional diaphragms. The thickness profile was found by interferometrically measuring the step height h at sharp edges of the layers. If the vapor profile was assumed to follow a function of the type

$$f(\varphi) = \cos^k \varphi \qquad (98)$$

(φ - angle, k - constant), and if the dependence of the vapor density on the relative tilt of the substrates was taken into account, the experimental data could be well fitted at $k = 0.95$ (normalized plot in Fig. 45). As it was found by simulations and experiments, the sensitivity of the final layer distributions against deviations of k is relatively small so that this value was further used in computations.

For the deposition of particular types of coatings (e.g. adhesion layers for polymeric substrates), additional techniques like sputtering had to be applied.

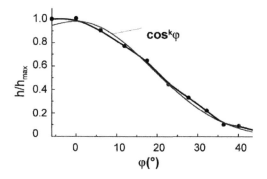

Fig. 45. Measured angular distribution of a vapor beam (fused silica) in comparison to a theoretical fit curve (exponent k = 0.95).

4.5.2. Types of shadow masks for the deposition of thin-film microstructures

The suitability of following *basic types of masks* for the deposition of thin-film microoptical structures (schematically in Fig. 46) was studied extensively:

(a) hole masks

(b) slit masks

(c) mesh-shaped masks

(d) wire-grid masks

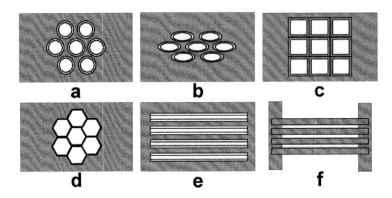

Fig. 46. Basic types of shadow masks used for the deposition of thin-film microoptical structures (schematically): (a, b) hole masks, (c, d) mesh-shaped masks, (e) slit masks, (f) wire grid masks.

Hole masks are small plates with regularly distributed small apertures of different shape (circular, elliptical, rectangular, square-shaped, hexagonal). *Slit masks* consist of linear arrays of small slits (length >> width). *Mesh-shaped masks* are hole masks with small structures between the holes (i.e., the holes fill most of the area). Highly precise mesh-masks can be fabricated with a (time-consuming) LIGA-like technology (structured deposition of metal, separation and transfer) [1999_2]. *Wire-grid masks* are comparable to slit masks and consist of an array of cylindrical shading bars (wires, fibers) or a single piece (coiled wire). Variable parameters are the distance to the substrate, minimal and maximal free hole diameter and thickness. In the case of 3D masks, the possible variation of the depth profile makes the design very flexible. In praxi, the technological problems and costs for low numbers of sophisticated masks are limiting factors. For many applications, however, the use of relatively simple mask geometries has proven to be a cost-effective and sufficiently precise alternative. In particular, apertures of optimized depth profiles could be well approximated by *linear (conical) depth profiles*.

It should be noted that the optimization procedure for masks with linear depth profiles is similar to the design problem for *mask-supported diffusion and ion-exchange processes* for the fabrication of GRIN structures where phase profiles are realized by index gradients instead of a thickness modulation. The similarity appears also in substructures of the corresponding distribution functions. Both types of microlenses tend to characteristic spherical aberrations by Gaussian-like distributions which either have to be reduced (as in the case of imaging microlenses) or exploited (for axial beam shaping).

For the fabrication of high-quality microlens arrays by thin-film deposition, two essential problems have typically to be solved simultaneously: (a) the fill factor has to be high, and (b) the elements have to approximate a parabolic shape as good as possible. Often, both demands behave in the manner of an uncertainty relation. With a careful control of mask and process parameters, however, two-dimensional, highly filled arrays of parabolic elements can be obtained as will be shown following.

4.5.3. Generation of arrays of high fill factors with the method of crossed deposition

In Chapter 2 it was already mentioned that high fill factors can be obtained by sequentially write cylindrical microstructures (crossed interaction zones). This principle formerly was only known from subtractive and modifying techniques. In simulations and test experiments, the transferability of this approach to the additive method of a mask-supported vapor deposition of microlenses with slit array masks was checked [1996_4, 1997_5, 1998_17, 1999_7, P3]. The 2D-superposition of optimized cylindrical microlenses was simulated on the basis of realistic deposition and mask parameters. It was found that it is possible to generate *closely packed arrays* with fill factors near 1 by depositing in 2 or 3 steps after rotating the substrates by angles of 90°, 60° or 30°, respectively. Grey-scale coded plots of the expected thickness distributions are shown in Figs. 47a,b. The rotation by 30° enables to fabricate elliptical (*anamorphotic*) microlenses Fig. 47c. The crossed deposition method could be convincingly demonstrated by deposition experiments with fused silica [1997_5, 1998_17], see Fig. 48.

Another advantage of wire-grid masks is the possibility to integrate *elements of different diameter*. In this way, nonuniform arrays with envelope thickness functions can be fabricated without bending the mask or additional macroscopic shadow masks. Improvements are possible by combining the crossed deposition with *intermediate etching*, where higher total layer thicknesses can be obtained, for the price of an increasing number of process steps.

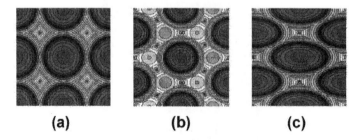

(a) **(b)** **(c)**

Fig. 47. Array structures resulting from a crossed deposition of linear arrays of cylindrical elements (lines of equal thickness representing theoretically calculated thickness maps): (a) orthogonal array of circular elements (2 steps, 90°), (b) hexagonal array of circular elements (3 steps, 60°), (c) hexagonal array of elliptical elements (2 steps, 30°).

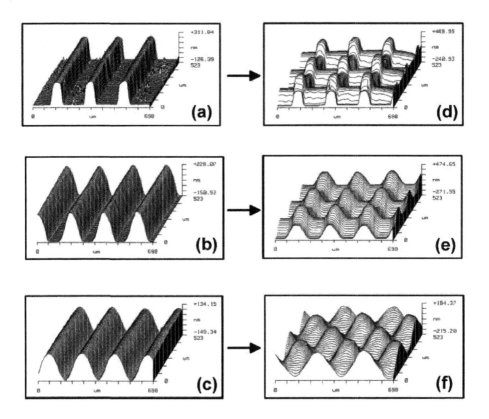

Fig. 48. Experimental prove of the method of two-step crossed deposition: Array structures resulting from a crossed deposition of linear arrays of cylindrical microlenses of different shapes and fill factors at varying distances of the slit array mask, left side: uncrossed, right side: crossed layers (fused silica layers on BK7 substrates, period 200 µm, slit lengths 3 mm).

Fig. 49. Working principle of crossed deposition of thin-film microlenses visualized by the specific focusing properties of different zones of the component (layer: fused silica on BK7 glass substrate, period 200 µm, thickness profiles corresponding to Figs. 48 c,f).

The working principle of the method of crossed deposition can clearly be visualized by the *imaging properties* of the different layer zones (Fig. 49). In the picture, the focal lines belong to the cylindrical parts, whereas circular focal spots are shaped by the crossed, spherical regions (left below). Because of the low thickness of the layer (< 400 nm), the focal length was relatively large (cm-range).

4.5.4. Wire-grid masks

The mask shapes with the highest design flexibility are three-dimensional shading structures with *curved walls* instead of linear ones. In general, such masks are of the highest possible complexity and price. However, there is a simple and easy-to-realize special case of slit masks: the *wire-grid mask* consisting of circular-cylindrical bars (schematically in Fig. 50).

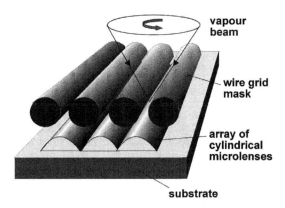

Fig. 50. Wire-grid mask consisting of an array of circular-shaped cylindrical bars as shading elements (schematically).

At very small diameters of the bars (thin wires), a parallel stretching of arrays of wires causes considerable technological problems (see Fig. 51). For this reason, wire-grid masks are preferentially interesting for the generation of relatively large microstructures where masks with stiff parts can be applied. In this case, the particular properties of wire-grid masks can be brought to bear. The first advantage is the capability to approximate smooth and nearly *parabolic profiles* under realistic deposition geometries with a very simple shadow mask structure, if the mask works in close contact to the substrate.

For the simulation [2000_1], the mask has to be separated into vertical slices of sufficiently small thickness (typically about 10). The contributions of all these

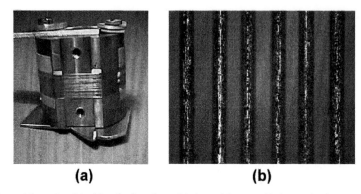

(a)   **(b)**

Fig. 51. Wire-grid mask with thin shading bars fabricated by a coiled metal wire: (a) complete device, (b) part of the wire grid (Ni wire, thickness 50 μm, period 100 μm). The photograph shows the problem of varying distances between the bars by mechanical deformations.

particular masks are calculated, superimposed and the resulting thickness profile finally normalized again. Thus, the evolution of the thickness profile as a function of the cycles of rotation could be simulated as shown in Fig. 52. The ratio of pitch (distance between the central positions of the shading cylindrical bars) to the diameter of the bars and the distance between substrate and bars were varied to optimize the conditions. The minimum deviation from a parabolic target profile was found in the case of direct contact of the wire-grid mask to the substrate.

The value of $x/r = 1$ in Fig. 52 corresponds to a fill factor of $\eta = 0.5$ (r is the radius of the bars, $p = 4r$ the period and $d = 2r$ the distance between the bars). This means that

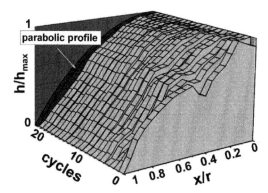

Fig. 52. Approximation of a parabolic target profile as a function of the number of cycles in a deposition system with planetary rotation and a wire-grid shadow mask in direct contact to the substrate (simulation). After about 10 rotations, the profile begins to stabilize for the used parameters [2000_1].

a fill factor of $\eta = 1$ can only be obtained by two sequential deposition steps in each spatial direction after a translation of the mask by half a period. The convergence of the profile over the deposition time was analyzed in theoretical calculations as a function of the radial position (Fig. 53). One can recognize that in the beginning the relative local thickness strongly oscillates with the number of cycles with different amplitudes and damping factors. Finally, the curves approximate a more or less steady state (< 10% maximum amplitude after 20 rotations). The calculations were found to be in relative good agreement with experimental results.

It has to be mentioned that crossed deposition has also been demonstrated with wire-grid-masks [1999_7]. (However, the number of necessary steps to get a high fill factor was 4 in this case so that this method is time consuming).

The use of shadow masks consisting of only a *single cylindrical element* for the shaping of convex layer profiles will be demonstrated later in the frame of graded AR-coatings where such kinds of profiles are required for particular applications. Arrays with elliptical elements can also be generated by depositing through slit array masks of different slit diameters as well as by changing the layer thickness from step to step as proposed in [P3] and also verified by experiments [1996_4].

A disadvantage of the method is the higher final layer thickness (without obtaining a higher NA) compared to single-step deposition. Another problem is the higher sensitivity against asymmetries of the profile function because the thickness errors of the particular process steps are added up.

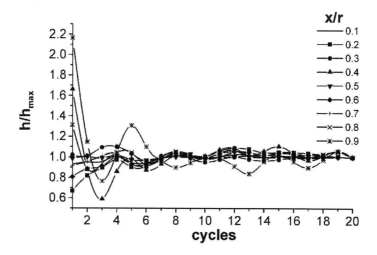

Fig. 53: Convergence of the microlens profile in Fig. 51 as a function of the number of rotational cycles analyzed at different axial positions (normalized) [2000_1].

4.5.5. Deposition of nonspherical elements with slit and hole array masks of conical apertures

For the deposition of thin-film microoptical elements of preferentially nonspherical shape, conical hole array and slit array masks were used by the author. Fig. 54 shows special types of metal masks which were fabricated by etching (Fig. 54 a, b) and laser processing with an Nd:YAG laser beam (c). The lower limit for the fabrication of masks of sufficiently good quality of the walls was found to be about 100 µm slit diameter (at a laser spot size of 30 µm).

Wall angles between 2.5 and 8° were obtained. In the case of commercially etched masks, maximum slope angles of the walls of 37.6° were available. Furthermore, lithographically etched masks of high structure precision and mechanically drilled masks were tested (not shown in the picture). The maximum number of holes in a mask was > 10^5 (mask type in Fig. 54b). Here, the dole diameter was 33 µm x 15µm and the array period 40 µm.

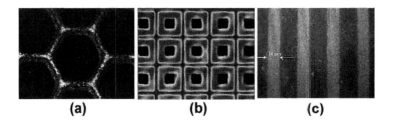

(a)   **(b)**   **(c)**

Fig. 54. Hole and slit array structures as shadow masks for the deposition of circular and cylindrical thin-film microoptical elements: (a) hexagonal hole array mask ("honeycomb structure", period 609 µm), (b) orthogonal array (period 40 µm, originally used as chromatography mask), (c) laser-cut metal mask (slit diameter 114 µm).

In Fig. 55, simulated distribution functions for cylindrical lenses of *nonspherical* shape in dependence on the distance between mask and substrate are depicted. The smaller diameter of the conical slit was directed to the substrate and the conical angle was large. The simulation was driven until the distribution converged (no further significant profile change). The minimum number of rotations for the computation was 20 [2001_1].

The resulting profiles can be well described by a set of *Gaussian and super-Gaussian functions*. Simulated profiles for another typical case are shown in the next picture (Fig. 56). Here, the larger diameter of the conical slit was directed to the substrate. The distribution functions are characterized by a Gaussian-like rim zone and a nearly linear slope from a (weak) kink to the center (prismatic central region).

Obviously, the profiles represent a superposition of two shading processes corresponding to two different parts of the mask.

In general it was found that the vapor deposition with shading 3D masks in systems with planetary rotation enables to shape *specific classes of distribution functions* with

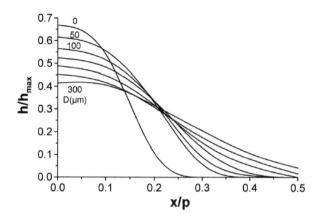

Fig. 55. Simulated shaping of cylindrical microstructures with Gaussian and super-Gaussian thickness profiles in one spatial direction by using slit array shadow masks with conical apertures (x - transversal coordinate, p - array period, small diameter directed to the surface of the substrate, parameter: distance between mask and substrate, steps of 50 μm, thickness normalized).

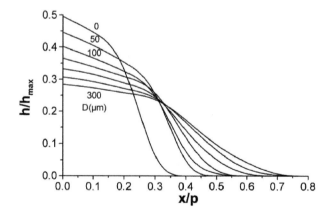

Fig. 56. Simulated shaping of cylindrical microstructures with nonspherical (peaked) profiles in one spatial direction by using slit array shadow masks with conical apertures (x - transversal coordinate, p - array period, small diameter directed away from the surface of the substrate, parameter: distance between mask and substrate, steps of 50 μm, thickness normalized).

Gaussian functions as an important subset. The thickness profile of this class of layers can be represented by the expression [2004_11]

$$h(x) = A \cdot |x| - B \cdot \exp\left[-C \cdot \left(\frac{x}{x_0}\right)^k\right] - D \qquad (99)$$

(A, B, C, D - parameters, x - transversal coordinate, x_0 - waist parameter, k - super-Gaussian exponent) and comprises (convex) Gaussian, super-Gaussian and conical functions and their inverse (concave) counterparts. Single layer structures of Gaussian or inverse-Gaussian shape act as *micro-axicon lenses* or (with HR coatings) as *micro-axicon mirrors*.

More complex distribution functions (e.g. as shown in Fig. 56) can be described by superimposing two or more functions of this type with each other or with additional parabolic functions. The interferometrically measured cylindrical thickness profiles of fused silica structures in Figs. 57 a and b are directly related to both of the above described cases.

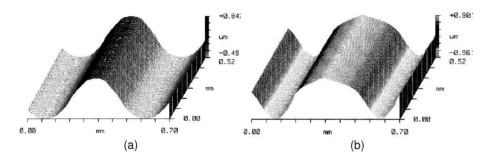

(a)   (b)

Fig. 57. Comparison of two thin-film micro-structures of different effective fill factors and shape functions deposited with slit-array shadow masks (SiO$_2$ layers, period 500 µm, measured with ZYGO^{TM} white-light interferometer). The profiles (a) and (b) correspond to the simulated cases (Gaussian-shaped, peaked) from Figs. 55 and 56, respectively.

Under optimized conditions, spherical central parts can be observed even with simple conical slit masks (Fig. 58). For applications where only the center of these cylindrical lenses is illuminated, the optical transfer functions can be relatively good. If the distance between mask and substrate is too large, it comes to an *overlap* of the outer parts of the structures. This effect can either support the profile optimization by re-shaping the distribution or it can lead to unwanted wings or valleys in the profile. The dented structures in Fig. 59a,b were obtained in just this deposition mode after single-step and crossed deposition. The elements in Fig. 59b have a concave central part.

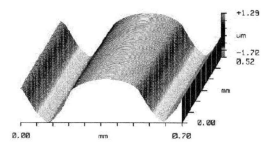

Fig. 58. Thin-film micro-structure with spherically shaped central region deposited with slit-array shadow mask (SiO$_2$ layer, period 500 μm, measured with *ZYGO*^{TM} white-light interferometer).

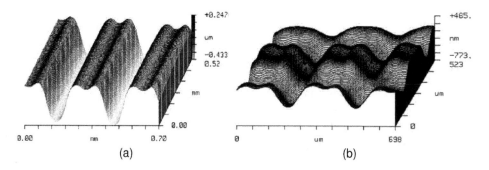

Fig. 59. Thin-film micro-structure deposited with a slit-array shadow mask at large distance, i.e. with overlapping rims: (a) structure after 1-step deposition with line-shaped central valleys in one direction, (b) structure after 2-step crossed deposition (angular step 90°) with quasi-square-shaped central conical parts (SiO$_2$ layer, period 500 μm, measured with *ZYGO*^{TM} white-light interferometer) [1998_17].

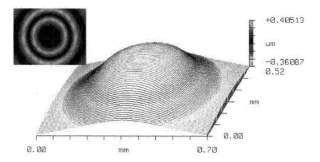

Fig. 60. Thickness profile of a single flat element from an array of microlenses (SiO$_2$ layer, thickness about 700 nm, measured with *ZYGO*^{TM} white-light interferometer). The inset shows synthetic fringes to visualize the excellent radial symmetry.

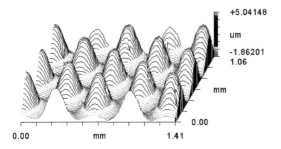

Fig. 61. Surface profile of a part of a hexagonal array of nonspherical micro-lenses (SiO$_2$ on silica, thickness about 6 μm, minimum period: 405 μm, total number of elements in the array: 1200, measured with *ZYGO*^{TM} white-light interferometer).

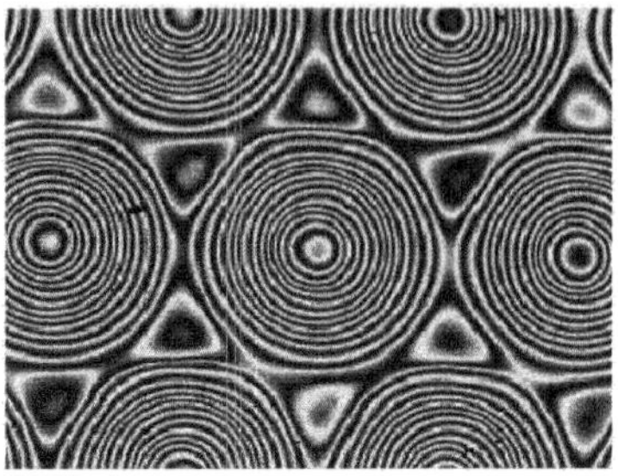

Fig. 62. Interferogram of a hexagonal thin-film microlens array of high fill factor (fused silica on a flexible polycarbonate substrate, period 300 μm, measured with *ZYGO*^{TM} white-light interferometer) [1998_17].

With hole masks, *radially symmetric structures* of profiles comparable to the above discussed types can be fabricated. The micrographs in Figs. 60 and 61 show a single Gaussian-shaped thin-film microlens of low thickness (about 700 nm sag height) and a part of a hexagonal array of thicker Gaussian-shaped elements (period 405 μm, thickness about 6 μm, total number of elements 1200). The fringe pattern in the interferogram in Fig. 62 demonstrates for a fused silica microlens array on polycarbonate (PC) how the maximum fill factor at least depends on the array geometry. In the hexagonal arrangement, the limiting factor is given by the star-shaped, uncoated area.

The generation of nonuniform arrays, i.e. arrays with variable thickness and/or shape function from element to element, was performed by using a second macroscopic mask (as in the case of GRMMA with envelope functions) and by bending or tilting

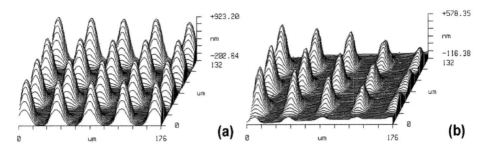

Fig. 63. Uniform (a) and nonuniform (b) arrays of small period ($p = 40$ μm, fused silica on silica, measured with *ZYGO*^{TM} white-light interferometer). The nonuniformity was reached by slightly tilting the shadow mask.

thin shadow masks (spatial variation of the distance mask-substrate). The latter case is illustrated in Fig. 63 for arrays consisting of relatively small elements. The array with the approximately linear envelope function for the maximum thickness h_{max} and slightly asymmetric (saw-tooth shaped) elements in Fig. 63b was obtained by tilting the mask by a small angle. It is compared to the uniform array without tilt in Fig. 63a.

4.5.6. Generation of arrays of high fill factors and parabolic profiles with mesh-shaped masks

Because of the specific limitations in the case of crossed deposition (higher base thickness of the layer) and slit array shadow masks (parabolic profiles contra high fill factor), alternative methods are needed to fabricate really closely packed parabolic thin-film microlenses of *high numerical apertures* (short focal lengths). One of the most promising approaches is the vapor deposition through optimized mesh-shaped masks [1999_2].

The principle of the mesh-masks was invented accidentally by using an available thin metal mesh as a shadow mask (a disused plasma electrode). The shape of a single element from the resulting hexagonal distribution can be seen in Fig. 64 [1998_17]. The picture shows that it is well possible to obtain both a nearly spherical distribution function and a fill factor close to 1 at the same time in only a single process step if a mask of very slim structures is applied.

This experimental finding encouraged to design and to test a new class of rather filigran masks. The main point was to find fabrication methods for masks with small bar diameter and techniques for positioning and mounting them [1999_2]. Details of mesh-shaped masks for the deposition of orthogonal arrays are shown in Figs. 65-66. In Fig. 65a, a part of an *orthogonal mesh-shaped mask structure* with a period of exactly 300 μm is shown. The masks were fabricated at the former Adolf Slaby Institute (now

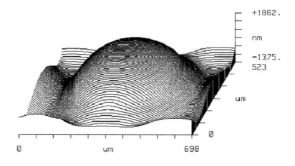

Fig. 64. Interferogram of a hexagonal thin-film microlens array of high fill factor (fused silica on silica, period about 600 μm, measured with *ZYGO*^{TM} white-light interferometer) [1998_17].

ASI Advanced Semiconductor Instruments GmbH). The total size of the masks was 49 mm x 49 mm and the number of holes 120 x 120 = 14400. Bar diameters between 6.8 and 54 µm were produced. At the crossing points, the mask bars had thicknesses between 9.2 and 54 µm. To improve the stability of the masks, frames of 4 - 47 µm thickness were integrated at the rim of the masks. The masks were either held at a distance to the substrate or fixed in direct contact with a permanent magnetic field. For this case, spacers of 2-4 mm height were integrated. It was found that a strong magnetic field causes a high risk of destruction so that distance holders have to be preferred. At optimized parameters, arrays for optical processors were fabricated (Fig. 66). The pictures show the realization of highly filled arrays of near parabolic elements with relatively sophisticated masks. However, the fabrication costs could be reduced by applying additional replication techniques.

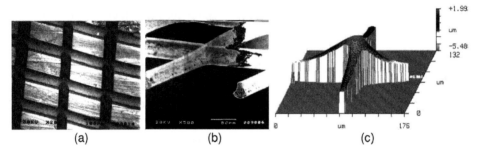

(a)   (b)   (c)

Fig. 65. Mesh-shaped masks for the deposition of orthogonal arrays of parabolic thin-film microlenses of high-fill factor: (a) top view on a part of a free-standing mesh-mask (Ni, period 300 µm, detected with a scanning electron microscope), (b) non-circular cross-section of a bar of the mesh, (c) integrated spacer of a mesh-mask (view from the bottom side which is in contact to the substrate during the deposition).

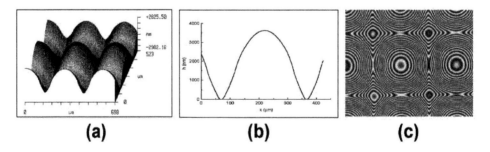

(a)   **(b)**   **(c)**

Fig. 66. Orthogonal thin-film microlens arrays of approximately parabolic shape with high fill factors deposited through mesh-shaped masks: (a) surface profile of a part of a microlens array (fused silica on silica substrate, period 300 µm, layer thickness about 3.5 µm), (b) thickness profile of a single microlens, (c) interferometric fringes (measured with $ZYGO^{TM}$ white-light interferometer).

4.6. Pre-processing of polymer substrates for improving the adhesion of thin-film microoptics

Mechanically flexible microoptical structures can be fabricated on the basis of shadow-mask supported thin film deposition on *polymer substrates*. This was first demonstrated with fused silica microlenses and multilayer GRMMA on polycarbonate (PC) and polyimide (PI) [1995_2, 1996_2, 1998_17]. Polymer foils and plates with typical thicknesses between 50 and 1000 μm were coated. To exploit the specific advantages (bending, low weight), the adhesion must be improved by pre-processing. Here, the high degree of complexity of such a procedure will be demonstrated for a selected example (*adhesion of fused silica on PC*).

The untreated polymer surfaces have to be analyzed, modified in an appropriate manner (depending on the materials of substrate and layer) and finally to be characterized with respect to the resulting adhesion. Systematic studies were performed in collaboration with ACA (Institute for Applied Chemistry Adlershof, Berlin) at fused silica layers on PC from two different manufacturers, General Electrics (PC/GE) and Röhm (PC/R) [1998_17]. The surface energy was determined by contact angle measurements with water, diiodomethane, formamide and benzyl alcohol. Surface energies of the untreated polymer substrates and their disperse and polar contributions are shown in Tab. 1. Here, the polar contribution is the essential one for the adhesion.

specimen (untreated)	surface energy (mN/m)	disperse contribution (mN/m)	polar contribution (mN/m)
PC/GE	44.2	43.4	0.8
PC/R	45.4	44.7	0.7

Tab 1. Surface energy of the untreated polymer foils (from contact angle measurements).

To improve the adhesion of silicon dioxide lasers, it was self-evident to take chemical reactions with plasma-activated *silicon-organic compounds* into consideration. In glass fiber reinforced materials, silicon-organic compounds are used as adhesive primer which acts via a bi-functionality (specific interactions with the polymer as well as the glass or silica). For tests with fused silica thin-film microoptics, γ-aminopropyltriethoxysilane ($C_9H_{23}O_3NSi$) was chosen. The expected chemical interactions of this compound in untreated state are shown in Fig. 67.

Under conditions of an additional *plasma excitation in a silane atmosphere*, radicals, electrons and ions generate active centers in the silane molecules and at the PC-surface so that a further strengthening of adhesive anchorage can be expected. Therefore, plasma pre-processing of substrates in a silane atmosphere was studied. In a

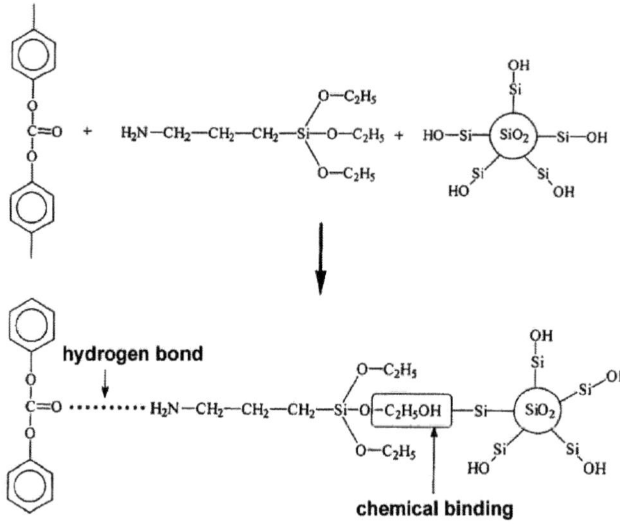

Fig. 67. Expected chemical interactions of γ-aminopropyltriethoxysilane with fused silica.

low-pressure microwave plasma device (producer: Saskia, Ilmenau, power 480 W, operating pressure 0.1 mbar, processing time 5-20 min), a *silanization* of the surfaces of the polymer foils was obtained. The high boiling temperature of 225°C at standard pressure and the enthalpy of evaporation of 11.8 kcal/mol required a heating of the silane to minimal 100°C corresponding to a vapor pressure of 20 mbar. The processing time was varied between 5 and 20 min. The initial and the final chemical compositions of the substrate surfaces were investigated with ESCA (electron spectroscopy for chemical analysis) in a spectrometer SAGE 100 (producer: SPECS).

After optimizing the plasma parameters, a significant enhancement of the surface energy was observed. The total surface energy after 10 min plasma treatment and their disperse and polar contributions are listed in Tab. 2. The comparison of the values for untreated (Tab. 1) and plasma treated substrates (Tab. 2) shows a significant increase of the polar contribution to the surface energy. Furthermore, the analysis of the chemical composition by ESCA indicated that the silane compound $C_9H_{23}O_3NSi$ was not only deposited on the polycarbonate surface but a part of the C_2H_5O- and H_2N-CH_2-CH_2-

specimen (untreated)	surface energy (mN/m)	disperse contribution (mN/m)	polar contribution (mN/m)
PC/GE	44.2	25.2	19.1
PC/R	58.1	34.6	23.6

Tab 2. Surface energy of the polymer foils after 10 min plasma treatment.

groups was already split off within the plasma. The increased percentage of Si and O in the structure promotes the deposition of SiO_2 microlenses.

The Improvement in adhesion by plasma pre-processing is also illustrated by the following experiment. On the basis of DIN 53530 standard, it was tried to determine the adhesion with the tape adhesion test. However, the adhesion force exceeded the destruction limit of the tape of (23.5 ± 1.8) N/cm. Therefore, this extremely high value has to be regarded as the lower limit for the adhesion. An exact measurement requires a further development of the method or an alternative technique. The mechanical flexibility of fused silica structures on PC is demonstrated in Fig. 68 by bending a component with bending radii down to about 2 cm.

Fig. 68. Mechanically flexible thin-film microoptical structures: fused silica microlenses on PC substrate (thickness 0.5 mm, diameter about 10 cm, thickness of fused silica layer < 8 µm, element diameter 720 µm, number of elements > 17,000) .

The *change of the focal distance* of cylindrical fused silica microlenses on polymer by bending flexible substrates was reported in [1996_2]. The focusing behavior of a single fused silica cylinder lens of a sag height of about 2.5 µm on a thin polyether sulphone (PES) plate (thickness 1 mm) is shown in Fig. 69. The interferometric fringes were detected in the straightened (Fig. 69a) and bended state (Fig. 69c). The bending curvature (radius = 99 mm) of the cylindrical lens in axial direction is drawn in Fig. 69e. Minimal and maximal focal distances were $f = 13$ mm and $f' = 8$ mm, respectively. The corresponding focal line images are given with Figs. 69 b and d. At smaller radii of curvature, destructions of the layer were observed.

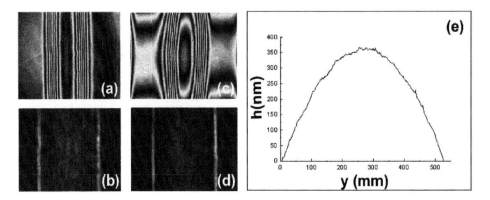

Fig. 69. Tuning the focal length of a cylindrical microlens between 13 and 8 mm by bending a flexible polymer substrate (lens: of a sag height about 2.5 μm, substrate: polyether sulphone). (a, c) Interferometric fringes, (b, d) focus image for the cases of straight substrate (a, b) and with 99 mm radius bent substrate (c, d). (e) Bending curvature of a cylindrical lens in axial direction.

By depositing arrays of circular microlenses, crossed cylindrical microlenses, anamorphic layers and multilayer structures on PC, PI and PES, the capability of polymer substrates for compact thin-film microoptical systems and adaptive-optical applications was demonstrated [1996_2].

4.7. Multilayer and compound microoptics

4.7.1. Multilayer microoptics

In the family tree of thin-film microoptics (see Fig. 43 in section 4.4.), one finds two examples for multilayer microoptical structures [2001_12], the *graded reflectance micro-mirror arrays* (GRMMA, Fig. 43g) and GRMMA with an additional *envelope* thickness function (Fig. 43h). For the fabrication of such components, the same method was used as in the case of microlens fabrication - the deposition through hole array shadow masks consisting of conical (3D) holes. Envelope functions can be obtained by an additional macroscopic shadow mask. This mask has to be fixed on top of the hole array mask and the substrate and rotates together with both. Figure 70 shows the thickness profile of a part of a micro-mirror array of super-Gaussian elements with a Gaussian envelope function measured with a diode based profilometer. The partially reflecting mirror consists of alternating HfO_2 and SiO_2 layers (design wavelength 1054 μm). In the center of the array, the maximum reflectance of the micro-mirrors was about 80%. Similar structures but with Gaussian profiles and with hard edges were fabricated as well [1993_4, 1993_5, 1995_2, 1995_3].

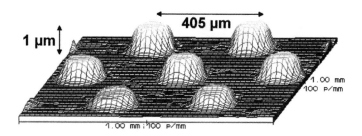

Fig. 70. Part of a micro-mirror array (period 405 µm) consisting of multilayer structures of super-Gaussian shape with Gaussian envelope thickness function measured with a diode-based profilometer.

Single multilayer micro-mirrors were designed as mode-selective elements in micro-disk lasers [1998_4, 1999_8]. By implementing these elements in a compact solid-state laser resonator, the stability range of the laser could be significantly extended (see paragraph 6.2.1.).

The deposition of *antireflection coatings* on top of dielectric microlenses is uncomplicated as far as the structure is flat enough (i.e. at low ratio of sag height to diameter). However, it can be a serious design problem if the contact angle is large (see above) or if the light source is strongly divergent. For this case, adapted coatings are necessary which we referred to as *graded antireflection coatings* [2001_1, P9]. Design, fabrication and characterization of such elements are demanding and will be discussed in detail in Chapters 5 and 6 in the frame of micro-reflectance measurement and laser beam shaping applications.

4.7.2. Metal-dielectric structures

(a) Refractive-reflective elements

The first *metal-dielectric microoptical structures* were developed as a simple alternative to the multilayer-type mode-selecting mirrors. Here, convex dielectric layer lenses were coated with a highly reflecting metal layer on the curved side so that they can work in reflective setup as a hollow mirror (Fig. 71a, [1998_4, 1999_8]). The disadvantage of this type of elements for many applications is that the laser beam has to pass the substrate twice so that it is reflected at 5 interfaces in a complete roundtrip.

For ultrashort-pulse laser beam shaping where a minimal dispersion is needed, another hybrid type of refractive-reflective microoptics was developed which consists of a metal layer coated with a thin dielectric layer [2003_1] (Figs. 71b, 72 and 73). The advantage of these refractive-reflective elements compared to the first mentioned type is that the light passes only (twice) through the dielectric layer but the substrate

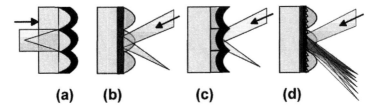

(a)   (b)   (c)   (d)

Fig. 71. Different types of hybrid metal-dielectric thin-film structures: (a) metal layer on convex dielectric microlenses (double pass through substrate and lenses), (b) convex dielectric microlenses on a plane metal layer (double pass through the dielectric layer only), (c) HR metal coating on a concave dielectric layer structure (purely reflective element), (d) convex dielectric microlenses on a diffractive metal grating (graxicon) (schematically).

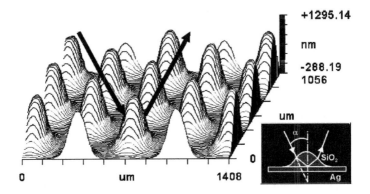

Fig. 72. Interferometrically measured thickness profile of a part of a hybrid refractive-reflective (pseudo-reflective) array corresponding to Fig. 71b. The structure was fabricated by depositing fused silica microaxicon lenses of Gaussian shape on a plane Ag mirror. The inset illustrates the working principle (double pass through the layer). The arrows symbolize the directions of incident and reflected light beams.

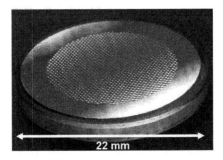

Fig. 73. Photograph of a hybrid refractive-reflective array for ultrashort-pulse laser applications (fused silica on Ag, corresponding to Fig. 72). The period is 405 μm, layer thickness about 1 μm.

dispersion is completely eliminated. On the other hand, one has to take into account that the structure suffers from parasitic Gires-Tournois resonances caused by the Fresnel reflection at the surface. Therefore, a further (broadband) reduction of the residual face reflection is useful. A second advantage follows also from working without a thick substrate. This brings the capability to operate at large angles of incidence which are typical for reflective setups in ultrafast applications as well as planar-optical design (zigzag paths by multiply folded imaging systems). It has to be noted that the phase is affected by an additional phase step at the interface between layer and metal.

(b) Purely reflective elements

Further progress in minimizing the dispersion was recently obtained by the introduction of purely reflecting thin-film structures (Fig. 71c, Fig. 74) [2004_15, 2005_6]. Parallel to the improvements in the design, the conical angle could be reduced by depositing extremely flat layers (< 1 µm). For ultrashort-pulse beam shaping applications, small conical angles reduce the travel time effects (geometric dispersion) as well. The interesting features of ultraflat axicons are described in Sections 4.9. and 6.3.

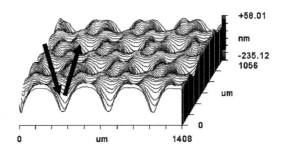

Fig. 74. Surface profile of an array of inverse-Gaussian shaped micro-axicon mirrors for ultrashort-pulse laser applications (Au on ZnSe). The structure was fabricated in two steps by depositing a dielectric layer through a shadow mask array followed by depositing a thin metal coating as broadband HR mirror.

(c) Reflective-diffractive elements

Very recently, new types of thin-film hybrid elements were proposed for *hyperspectral* applications which combine the spectral selectivity of diffractive gratings with the particular beam shaping properties of refractive thin-film microaxicon arrays generating axially extended focal zones (schematically in Fig. 71d) [2006_8, P16]. The elements were called "*graxicons*" in allusion to the so-called "*grisms*" [1970_2] (which represent the combination of prisms and gratings and are widely used in spectroscopy, microscopy, astronomy and compressor-stretcher systems). Two different types of graxicons are shown in Figs. 75 a and b, a pseudo-reflective version and a purely reflective version. In the schematic representation, the relations between thickness and transversal extension were deliberately blurred to better visualize the structure. If the pseudo-reflective arrangement (with a double pass of the light field through the axicons like in the case of the refractive-reflective elements in Fig. 71b) is replaced by the purely reflective case of the imaging grating (Fig. 75b), parasitic reflection and phase distortions are avoided.

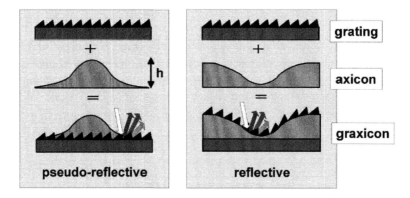

Fig. 75. Graxicons (schematically) as a combination of diffractive structures (metal gratings) and refractive microoptical arrays (convex dielectric microaxicons): (a) pseudo-reflective arrangement, (b) purely reflective arrangement. Typical scales for thickness and radial coordinate differ by a factor in the range of 1:1,000 so that the real elements are very flat.

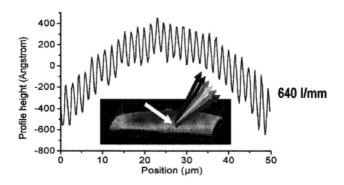

Fig. 76. Part of the surface profile of a pseudo-reflective graxicon (fused silica layer on a metal grating, 640 l/mm). The arrows in the inset with 3D projection symbolize the diffraction (profile measured interferometrically).

In a first experimental approximation, a *pseudo-reflective graxicon* was realized by depositing a fused silica microaxicon array on a reflective metal grating of a line density of 640 l/mm (Fig. 76). *Purely reflective graxicons* (design in Fig. 75b) are currently under investigation. In principle, the existing technologies allow to fabricate such elements. The key problem, however, is to find cost-effective methods and/or replication techniques.

Graxicons enable for a spatio-spectral mapping of light fields. Therefore, they are promising candidates for advanced types of wavefront sensors and optical processors (see paragraph 7.4).

4.8. Structure transfer of thin-film microoptical components by reactive ion etching

The central problem of microoptical hardware is to transfer stepped or continuous profiles from a mask or initial pattern into the surface or volume of interest [1997_39]. In paragraph 2.8 it was mentioned that microoptical profiles written in a first material can be transferred into layers or substrates consisting of identical or different materials. A wide-spread subtractive transfer technique is based on reactive ion etching (RIE) [1993_2]. Typically, this procedure is applied to photoresist structures. With the same technique, however, thickness-modulated dielectric layers from mask-shaded deposition steps can further be transferred.

Transferring the structures into bulk material or replicating them can lead to a loss of specific advantages of thin-film microoptics. On the other hand, the etching selectivity in a transfer process can be used to obtain extraordinarily flat but well-defined continuous-relief structures of low roughness. Such structures are difficult to fabricate by other known technologies like mechanical treatment, thermal reflow or photolithography. For example, fused silica microlenses of low numerical apertures were produced by exploiting the etching selectivity in the transfer from a resist structure [2001_3].

If the process works sufficiently linearly, the thickness modulation transfer is directly proportional to the etching selectivity S (i.e. the ratio of the etching rates R_2 and R_1 of bulk and layer materials). After transferring a structure of the initial height $h(x)$ at a spatial position x, the final structure height $h'(x)$ can be expressed by the simple formula:

$$h'(x) = \frac{R_2}{R_1} \cdot h(x) = S \cdot h(x) \ . \tag{100}$$

The RIE transfer from thin dielectric layers (period 40 µm, thickness modulation depth 147 nm) into dielectric substrates of identical material with an excellent conservation of

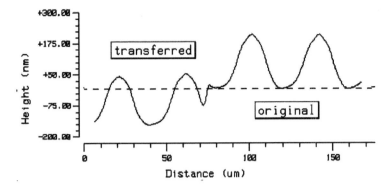

Fig. 77. Transfer of refractive fused silica microstructures into a silica substrate by reactive ion etching. The comparison of initial and final shape function shows an excellent agreement.

the shape function was demonstrated experimentally [1998_17] (Fig. 77). The method was extended to transfer layer structures from fused silica and hafnium dioxide into non-identical substrate materials (silica, calcium fluoride), preferentially for VUV laser applications at wavelengths down to the fluorine laser wavelength (157 nm [2004_1, 2004_2]. The intermediate steps of structure transfer were analyzed interferometrically over a time interval of in total 45 min (Fig. 78). After 20 - 40 min, the most unstable state was observed where the shape and diameter were extremely different from the initial ones. At the final stage, the structure parameters return relatively fast to the initial values (last 5 min). FWHM diameters of the convex Gaussian-shaped microstructures of 180, 157, 135 and 169 μm were found after 0, 5, 20 and 45 min of etching time, respectively.

Figure 79 conveys a further impression of the RIE process: The line between the untreated and etched surface region of a specimen a few minutes after process start runs about diagonally through the field of view.

By taking advantage of an etching selectivity of $S = 20$ for the transfer from SiO_2 to CaF_2, a strong flattening effect was obtained ($h_{max} = 2\mu m$, $h_{max}' = 0.1$ μm). The rms roughness changed from initially 2.3 nm to a final value of 2.6 nm (corresponding to 0.015 λ and 0.017 λ at a wavelength of $\lambda = 157$ nm). To compare, the roughness of the unprocessed substrate was about 1.8 nm (0.011 λ). The minimum thickness modulation depth reached by this method up to now was 60 nm. This corresponds to an average conical ray angle of about 0.015°.

The second main advantage of transfer into bulk material is that the final structure is more robust against mechanical and thermal stress because it is *monolithic*. Monolithic continuous relief components are especially suited as master structures for a replication.

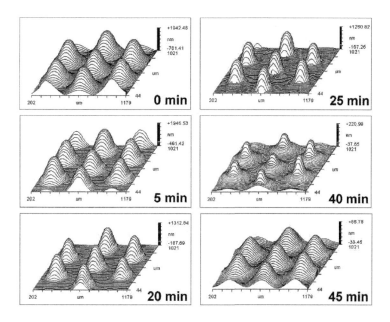

Fig. 78. Analysis of the intermediate states of the RIE transfer of fused silica into a calcium fluoride substrate. Parameter is the etching time. The etching selectivity was about $S = 20$ (CaF_2 material: Korth Kristalle).

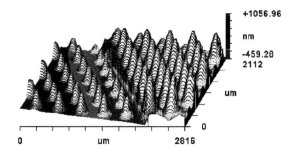

Fig. 79. Direct comparison of etched and unetched zones during the RIE transfer from fused silica into calcium fluoride (similar to the process see Fig. 78 but starting with lower thickness).

4.9. Nanolayer microoptics

The above described etching transfer is a first method that enables for the generation of really *ultraflat* (smallest-angle) microoptical structures. A second one is the *direct deposition* of extremely thin layers (referred to as *nanolayer microoptics* [2005_10]). The technical challenges of nanolayer deposition consist in the proper control of the deposition conditions (closed rotational cycles, homogeneity of the vapor distribution) at a relatively short deposition time as well as in the substrate quality (waviness, roughness, homogeneity of the refractive index). The minimum obtained structure depth was about 30 nm. At a wavelength λ = 150 nm this corresponds to $\lambda/5$. Therefore, the demands concerning the tolerable substrate roughness can no longer be expressed in terms of the *Rayleigh criterion* ($\lambda/4$) but have to be reduced even down below the *Maréchal criterion* ($\lambda/14$). For typical element diameters of 500 μm, a height of 30 nm corresponds to a *ratio between thickness modulation depth and diameter* as small as h/d = 7.5 x 10^{-5}. For spherical microlenses (calotte), the corresponding *radius of curvature* can be calculated from eq. (14) and is found to have the exorbitant value of 1042 m! By using eq. (97) one obtains a (theoretical) *contact angle* of only 0.01°. The *conical angles* are (depending on the refractive index and the shape function) comparably small. The angular distribution for a fitted Gaussian thickness profile of a CaF_2-microaxicon with a height of 55 nm and a FWHM of about 150 μm is shown in Fig. 80. (More details on the calculation of angular distributions of surface-relief-type microoptics can be found in paragraph 6.3.3.).

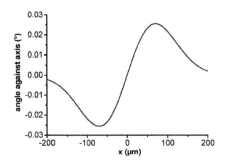

Fig. 80. Angular distribution of a Gaussian-shaped microaxicon of the height of 55 nm and a FWHM of the thickness distribution of about 150 μm (layer directed in propagation direction, plane wave illumination, wavelength 157 nm). The average conical angle (against the optical axis) is 0.015°.

The realistic examples show that the *scaling of microoptical structures* in depth dimension (not in transversal scale - as in the case of DOE and nanostructures) *down to the few-nanometer range* (i.e. well below the wavelength) opens a quite new and unconventional playground for the shaping of beams and ultrashort wavepackets. From point of view of diffraction, the obtainable long focal distances are of particular interest for *short-wavelength applications* (VUV, X-ray) because of the more extended range of nearly diffraction-free propagation (see the dependence of the Fresnel number on the wavelength).

For the theory of light propagation and diffraction, the now available experimental access to the other end of extreme parameters (see Fig. 83, next section) brings new challenges as well. Gaussian-shaped nanolayer microaxicons have a vanishing contact angle so that a sharp edge doesn't exist and one has to expect (self-) apodization effects caused by the phase profile. Related problems arise from ultraflat rim zones of waveguiding elements or graded thin multilayers (still to be understood). Interference effects have to be described on a low-contrast scale. And the optics of refraction (tailoring wavefronts far from diffraction) or the concept of "rays" (shaping needle-like beams of unprecedented extension) come closer to a physical realization.

Apart from these more general considerations, the specific features of nanolayer microoptics promise to have a large potential for practical applications in next generation measuring techniques and materials processing. First achievements towards these techniques are presented further down.

4.10. VUV-capable transparent thin-film microoptics

For multichannel lithography, patterned microstructuring, spectroscopy or a spatio-temporal control of short-wavelength sources like excimer lasers or high harmonics of Ti:sapphire lasers, ultraflat optical components with the capability to shape beam arrays at wavelengths down to $\lambda = 157$ nm or less are required. As a less time-consuming and more cost-effective alternative solution for the fabrication of VUV beam shapers compared to the transfer via selective etching, the direct deposition of VUV-transparent layers on substrates of identical material is very attractive. Recently, arrays of cylindrical and circular microlenses and microaxicons consisting of MgF_2 were obtained by directly depositing structured MgF_2 nanolayers onto MgF_2 substrates with

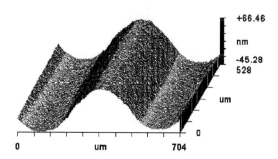

Fig. 81. Nanolayer microoptics: single cylindrical microaxicon of Gaussian cross-section (part of a linear array of 20 elements, period 500 μm, thickness modulation depth about 100 nm, MgF_2 layer on a plane MgF_2 substrate).

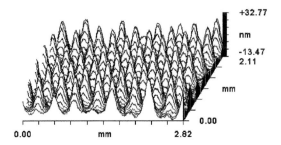

Fig. 82. Nanolayer microoptics: circular-shaped microaxicons of Gaussian cross-section (part of a hexagonal array of 1200 elements, thickness modulation depth about 100 nm, MgF$_2$ layer on a plane MgF$_2$ substrate).

the shadow mask technique [2005_9, 2005_10, 2006_6]. Figs. 81 and 82 show parts of arrays of cylindrical and a circular elements with thickness modulation depths of about 100 nm and 35 nm, respectively. The short-wavelength edge of the spectral transmission curve of the deposited MgF$_2$ structures begins at $\lambda = 120$ nm (50% transmission). At a wavelength of $\lambda = 157$ nm, the transmission reaches a value of about 85%. (To compare, the structures transferred into CaF$_2$ show 50% and 78% transmission at $\lambda = 130$ nm and $\lambda = 157$ nm, respectively).

It has to be mentioned that also *slightly absorbing layers* can be applied to beam shaping if the layers are thin enough (low absolute absorption), losses are not critical, and the power density is not exceedingly high (because of the damage risk). In VUV experiments, thin fused silica layers on calcium fluoride substrates were successfully applied. In paragraph 6.5.2. it will be shown that the absorption of layers can also be exploited to obtain a beam cleaning by *amplitude self-apodization*.

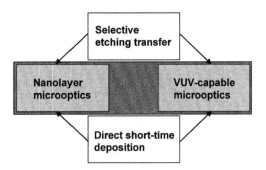

Fig. 83. Quintessence of the remarks on the fabrication methods for nanolayer and VUV-capable microoptics. Both selective etching transfer as direct deposition techniques enable for generating array structures of low thickness modulation depth. Both come with specific advantages and disadvantages so that one has to weight the odds in practical applications.

The scheme in Fig. 83 shows (without claiming to be complete in any way) the *extract* of the past three paragraphs. Both nanolayer microoptics as well as VUV-capable structures can be fabricated by selective etching transfer as well as by direct short-time mask shaded vapor deposition in a rotating system.

4.11. Problems and trends

As already pointed out, one of the most serious problems to be solved for thin-film microoptical fabrication technologies is still the *quality of the masks* (which is closely related to the deposition of material on the mask itself during the fabrication process). High-precision masks are realistic solutions only for basic research tasks or in the case of extraordinarily high demands on the element quality. As a base for a replication, classical microoptical master components can be more appropriate in many cases. For medium-level tolerances, cost-effective *one-way masks* (e.g. made of polymer) as well as effective mask cleaning and recycling might be interesting options. Improved techniques for the *manipulation of masks* (mounting, stretching, positioning, tilting etc.) are required.

Another open task is to systematically investigate the limits of generating (transversal) *nanostructures. Removable* resist masks and self-organized polymer structures are currently investigated as small shadow masks.

The coating of thin-film microlenses with *self-assembled small polymer spheres* (as anti-reflection structures in moth-eye design) can lead to new applications (e.g. reflectionless IR systems). In pilot experiments [1998_17], the area of regularly ordered latex particle structures without dislocations was typically limited to the range of several thousand elements. A part of such an ordered region of in total 60 x 60 μm^2

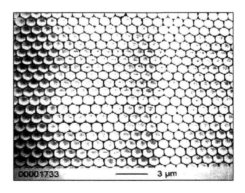

Fig. 84. Ordered region of an array of self-assembled polymer micro-spheres on the surface of fused silica thin-film structures (test experiments for moth-eye design).

corresponding to 2000 elements of 1.5 μm diameter on the surface of a fused silica microlens monitored with SEM is shown in Fig. 84. (For IR applications at wavelength > 4 μm, thin-film structures of ZnS and ZnSe can be deposited instead of fused silica test layers used in the experiments).

Additional functionality can be achieved by *integrating layers or substrates with particular properties*, like:

- nonlinear optical characteristics
- fluorescence
- spectral absorption or gain
- polarization
- scattering.

Currently, the combination of transparent and reflective thin-film microoptical arrays with nonlinear semiconductors (ZnO) and glass with incorporated semiconductor nanoparticles is under investigation [2003_10, 2004_7, 2005_13, 2005_33, 2006_25] (see also Chapter 7).

Transversal and 3D structural features of thin-film microoptical design can extend the functionality as well, e.g. by:

- post-processing (laser direct writing and etching, e.g. of grating structures, doping and index tailoring by ion-implantation, diffusion or photorefractive effect)
- structured growth (sculptured layers)
- 3D design for multilayers (nanocrystals)
- phase-shaping multilayer design
- reflective X-ray thin-film microoptics
- polarization and orbital momentum management with structured thin-films.

Hybrid systems integrating conventional and thin-film microoptics in compact systems are promising approaches (see Chapter 6). A contrary, but not less interesting approach is the application of shadow mask techniques for *smoothing surfaces*. Recently, the surface of decimeter-scale lithium niobate for solar observations was flattened down to a rms-roughness < 0.3 nm [2006_2].

Chapter 5

CHARACTERIZATION OF THIN-FILM MICROOPTICS

5.1. Specific measuring problems

The major part of tasks for the characterization of thin-film microoptical components is well known either from the characterization of microoptics in general, or arises from the properties of thin optical layers or multilayers. The corresponding techniques are briefly mentioned in this paragraph. Furthermore the chapter focuses on a third group of techniques, which is related to specific features of flat thin-film micro-optical structures.

A first important point is the measurement of the surface profiles of single elements and arrays and related physical quantities like roughness, waviness or radii of curvature [1995_7; 2005_25]. It is helpful to separate relevant scales with respect to the corresponding spatial frequencies [1989_2]. For the shape measurement on a scale between 1 and 1000 µm, optical methods like interferometry and diode laser profiling with differential focus sensors as well as mechanical scanning methods (stylus) are used. The drawback of stylus technique is the risk of scratching. Smaller features on a scale < 1 µm can be inspected with scanning electron microscopes (SEM), atomic force microscopy (AFM) and scanning near-field optical microscopy (SNOM). High-resolution measurements are necessary to evaluate scattering processes and to detect substructures of layers on the nanoscale. The comparison between the results obtained on different scales is all but trivial because of the different depth resolution and other peculiarities of the used measuring techniques (see also [1992_4]).

Interferometric methods (Mach-Zehnder interferometer [2001_4, pp. 141-163; 2005_34], Twyman-Green interferometer [1995_33, 2005_34, 2005_35], Michelson interferometer, Fizeau interferometer [2001_4, pp. 132-133], Mirau interferometer [1998_17], Zeiss Linnik interference microscope [1997_2, pp. 192-193]) enable to resolve minimal thickness variations. With phase-shifting procedures [1993_13] and in multi-wavelength operation modes, surface profiles and small steps can be resolved even in Ångström range. Commercial systems are endued with software tools for optical functions like modulation transfer function (MTF), point spread function (PSF), Strehl ratio (SR), encircled energy (EE) or to perform the Fourier transform of a phase distribution. For details on interferometric setups for the characterization of microlenses and microlens arrays, we like to refer to the literature [1991_12, 1994_39,

2001_4, 2005_34, 2005_35] and for the technique of interferometry in general to the classical book of Malacara [1992_6]. A comprehensive overview on interferometric methods of microlens array testing one fin ds in [2004_1].

A comparable and versatile optical characterization method for microlenses is the *digital holographic microscopy* [2002_42, 2006_27]. An advantage of digital holography is to allow for a reconstruction of complete phase distributions in space from a single measurement and thus to shorten the detection time.

Characterization techniques of the second group, i.e. specific for thin optical films, multilayer structures and the control of film growth, are also well described in Refs. [1995_26, pp. 269-474; 2002_36, pp. 382-495; 2002_37, pp. 488-532; 2003_26; pp. 181-206]. They include the determination of refractive indices, optical constants, absorption coefficients, scattering as well as mechanical properties. In many cases, the spatial resolution or sensitivity of thin-film measuring techniques is not high enough to fully characterize *thin-film microoptics* and in particular arrays of large numbers of small elements. Only recently, commercial ellipsometric devices reached spatial resolutions comparable to optical microscopes. To characterize *multilayer microoptics*, the spatially variable reflectance or transmittance has to be detected. Because of the microscopic dimensions of the elements, however, a high spatial resolution is required. This can be obtained by using a focusing measuring beam. However, if the multilayer design is *spatially and angularly* adapted (e.g. in the case of graded AR-coatings, as it will be shown later on) the characterization method has to be highly resolving *in space and direction*. This is a new challenge. A further hard challenge arises from the task to do such measurements under the conditions of extremely weak signals (e.g. a residual reflection R_{res} of anti-reflection layers down to the range of 10^{-5}).

In the following we will concentrate on selected techniques which are of special interest for testing flat thin-film microoptical structures of the types described in Chapters 4, 6, and 7. In particular these are:

- interferometry of thin-film microlens arrays and array-specific data analysis, and
- the spatially and angularly resolved specular reflectance of reflecting and anti-reflection type multilayer microoptics.

5.2. Interferometric characterization of thin-film microlens arrays

5.2.1. Phase shift interferometer

Most of the thin-film structures used for the applications described in this book were characterized with a *ZYGO*™ Maxim*GP (LOT/Oriel) white-light phase shift interferometer with Mirau-type and Michelson type microscope objectives of 100x, 40x, 10x, 5x and 2.5x magnifications (Fig. 85). The interferometer was calibrated with a highly precise reference plane (crystal cleavage plane). The measuring wavelengths were $\lambda_1 = 616$ nm and $\lambda_2 = 645$ nm.

The schematic of the setup is shown in Fig. 86. A white-light source is collimated, spectrally filtered and passes a beam splitter. The collimated beam is reflected at a partial mirror and focused by an objective onto the microoptics to be characterized. A piezo-translator changes the position of the objective thus vertically scanning through the structure. The beam reflected from the spot on the surface is imaged onto the CCD camera where the first reflected and the object beam interfere. Over the coherence length of the light beam (a few micrometers), correlograms are obtained with sufficiently high fringe contrast to be processed by a fringe processor software.

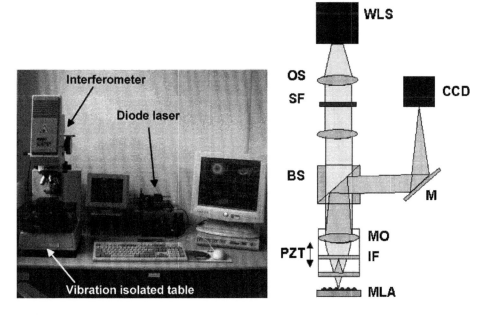

Fig. 85. Measuring station for the optical characterization of thin-film microoptical components based on a *ZYGO*™ interferometer for surface profilometry and a fiber-homogenized, collimated diode laser (635 nm) for propagation measurements.

Fig. 86. Schematic of the interferometer setup with Mirau measuring head. (WLS - white-light source, OS - optical system for the collimation, SF - spectral filter, BS - beam splitter, CCD - matrix camera, PZT - piezo-translator, MO - microscope objective, IF - interferometer head, here Mirau-type.)

5.2.2. Characterization of shape distribution and periodicity of microlens arrays

The typical data processing procedure is now demonstrated by the example of the parabolic microlens array shown in Figs. 66a-c (see paragraph 4.5.5.). In this case, parasitic reflections were minimal because of nearly identical refractive indices between substrate and layer (fused silica on silica).

Beside the shape function of single elements and sag height distribution, the *uniformity* of the array was characterized by a statistical analysis. At first, the rms (root mean square) was determined to be 3 nm. Mathematically, the rms represents the standard deviation (which is the square root of the second statistical moment or variance) [2006_22]. In the data analysis, the difficulty is to properly separate the rms from the shape function what can be done by Fourier filtering (but suffers from residual errors in case of overlapping frequency spectra.)

Width and shape of the *amplitude spectrum* (Fig. 87) deliver information on the shape variance of the single elements and the periodicity of the array. It was found that spatial frequencies < 10 mm^{-1} dominate and the distinct peak corresponds to the period of the array ($p = 300$ µm).

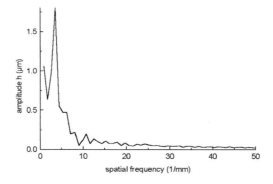

Fig. 87. Amplitude spectrum of the spatial frequencies of the amplitude distribution corresponding to the orthogonal thin-film microlens array structure shown in Figs. 66a-c [1999_5].

The *autocovariance function* (Fig. 88) describes the covariance of the distribution with itself and thus can be used to evaluate the quality of periodical structures like arrays [1999_5]. It represents the correlation of the microlens profile with a replica of itself which is obtained by *lateral shifting*. For a profile of finite length and a discrete set of N data points z_i measured at distances τ, called "lag lengths",

$$\tau = l \times \tau_0 , \tag{101}$$

(l - integer number, τ_0 - step width), the autocovariance function $G(l)$ can be defined as [1989_2, p. 44]

$$G(l) = 1/N \times \sum_{i=1}^{n-l} z_i \times z_{i+l} \qquad (l = 0, 1, 2, ...N\text{-}1) . \tag{102}$$

(Note that τ here is a spatial parameter and not a temporal one. This nomenclature was adopted to stay in accordance with the literature.) In the special case of vanishing lag length, $\tau = 0$, the dimensionless autocovariance function is closely related to the square of the rms roughness of the profile (second moment or variance). The particular lag length τ_c which corresponds to a reduction by 1/e is defined as the *correlation length* of the autocovariance function:

$$\tau_c = G(0)/e \ . \tag{103}$$

For the lens data from Figs. 66a-c, a correlation length of about 70 μm was determined.

The *Fourier transform* of the autocovariance function delivers with the *power spectral density* further information on the structural features of the array of interest. The density plot of the 2D-Fourier spectrum of the same microlens array in Fig. 89 indicates a relatively high degree of symmetry despite of some small differences between x- and y-direction (i.e. the directions parallel to the edges of the square-shaped elements of the array).

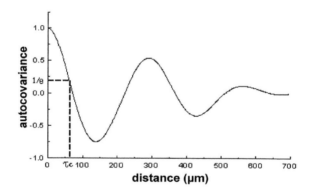

Fig. 88. Autocovariance function $G(\tau)/G(0)$ for the spatial frequencies of a part of the orthogonal thin-film microlens array with $G(0)=900\text{Å}^2$ (τ_c - correlation length, data according to Figs. 66a-c and Fig. 87, [1999_5])

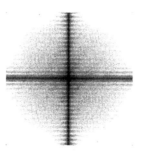

Fig. 89. Density plot of the 2D Fourier spectrum of surface spatial frequencies for a part of an orthogonal thin-film microlens array (data according to Fig,. 66a-c, [1999_5]).

5.2.3. Optical functions of thin-film microlens arrays

On laboratory stage, manifold innovative methods for the generation of microoptical structures are tested worldwide and a flood of papers on physical and chemical procedures can be found in journals. It is however striking that some of these papers end their characterization with the presentation of some rough stylus data (even limited to one direction if there is obviously a spatial asymmetry). They claim "high fill factors" for at least strongly aberrated lenses where the fill factor is not a measure for the optical efficiency. On the other side, some people involved in measuring techniques like to present the best-shot data obtained with nice and easy-to-handle calibration objects of moderate parameters instead of speaking about error bars.

On the basis of interferometrically measured height distributions and optical design and propagation software, however, a reliable analysis of optical functions is possible in many cases. To illustrate this, we continue with our example from Figs. 66a-c. The focusing properties were simulated by a beam propagation method (BPM) [1992_6, 1995_34, 1995_5] for one-dimensional coherent illumination. The BPM program allows for linear as well as logarithmic plots of phase and intensity, a normalization for each axial coordinate and cuts. Propagation experiments were performed by illuminating the arrays with a collimated laser diode ($\lambda = 635$ nm) and monitoring the intensity distribution in the plane of interest with a microscope with CCD camera or the interferometer setup itself. The angle of incidence was changed by rotating the array. Intensity profiles were also detected for negative distances (planes behind the lenses) to identify concave areas of the arrays (e.g. in the space between neighboring lenses).

The light propagation behind 5 elements of the array simulated with BPM is shown in Fig. 90 in three different modes of representation to visualize different features like (a) the focal zones, (b) the outer regions and (c) the behavior over the complete propagation path.

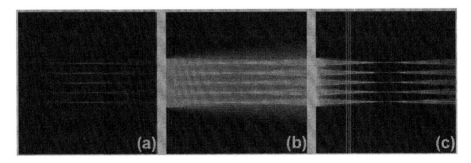

Fig. 90. Focusing properties of an array of 5 microlenses simulated with BPM on the base of the data from Figs. 66a-c (period 300 μm, interval in horizontal direction: 10 mm, wavelength 635 nm), (a) linear, (b) logarithmic, (c) normalized on local intensity maxima [1999_5].

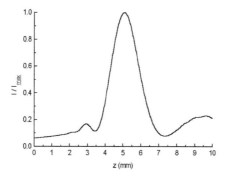

Fig. 91. Axial intensity distribution along the focus of a single microlens (simulation with BPM, linear cut in axial direction, thickness profile according to Figs. 66a-c, $\lambda = 635$ nm).

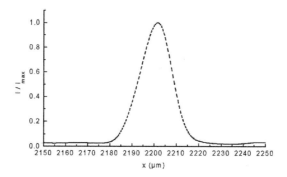

Fig. 92. Transversal intensity distribution in the focus of a single microlens calculated with BPM (simulation with BPM, linear cut in axial direction, thickness profile according to Figs. 66a-c, $\lambda = 635$ nm).

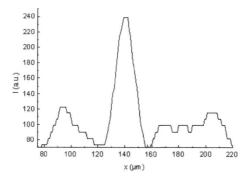

Fig. 93. Measured transversal intensity distribution in the focus of a single microlens (same structure as used for the simulations, thickness profile according to Figs. 66a-c). Compared to the theoretical intensity distribution, more energy dissipates into the side-lobes.

Calculated intensity profiles for axial and transversal cuts through the focal spot
(corresponding to the beam focused by the microlens at the central position) are
presented in Figs. 91 and 92. The spot diameter was theoretically estimated to be about
30 µm. This value agrees well with the experimentally measured diameter of the Airy
disk (Fig. 93). However, stronger side-lobes appear in the experiment. This discrepancy
is caused by the limitations of the model (1D, coherent) and illustrates the necessity of
two-dimensional calculations for a realistic evaluation of imaging properties.

To determine standard optical functions, the interferometric profile data were fitted
by Zernike polynomials [1995_27; 1998_2]. *Modulation transfer function* (MTF)
[1974_5, pp. 266-269; 1999_1, pp. 18-21; 2005_2, pp. 65-66; 2006_28, pp. 266-271],
point spread function (PSF) and *encircled energy* (EE) have been determined for
concentric, circular parts of single elements of the known microlens array (Figs. 94-96).

The dependence of the optical functions on the diameter of the synthetic apertures
in our realistic example shows convincingly that even *small deviations* from the
parabolic shape in the rim zone (spherical aberrations) and diffraction lead to a
smearing of optical information in space. In case of ultrashort pulses, equivalent
dissipation effects are observed in the temporal domain caused by dispersion effects.
The best imaging properties (minimum aberrations) correspond to the central part of
the thickness distribution as expected.

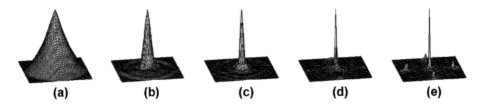

(a)   **(b)**   **(c)**   **(d)**   **(e)**

Fig. 94. Qualitative comparison of the modulation transfer function (MTF) for different circular,
concentric parts of varying radius of a single lens of the orthogonal microlens array. Data
correspond to Figs. 66a-c, reference wavelength here: 648 nm, abscissa proportional to optical
path difference (OPD) / radius. The diameters of the software masks were: (a) 50 mm, (b) 100
µm, (c) 150 µm, (d) 200 µm, (e) 250 µm.

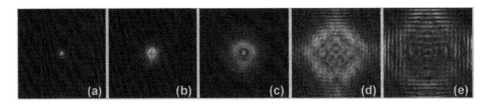

Fig. 95. Two-dimensional point spread functions for the element and value settings as in Fig. 94
(abscissa proportional to OPD / radius).

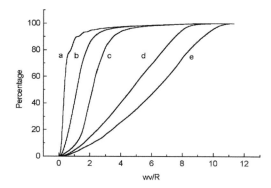

Fig. 96. Encircled energy (EE) for the element and value settings as in Fig. 94 (abscissa here: wave number / radius).

5.3. Reflectance mapping of multilayer microoptics

5.3.1. Spatially resolved measurement of specular reflectance

For the *spatially resolved* characterization of the specular reflection partially reflecting elements (e.g. GRMMA, semiconductor diode facets) and low-reflectance microoptical objects (e.g. micro-graded-AR-coatings), a double-stage lock-in technique was used (schematically in Fig. 97) [2000_5, 2000_13, 2001_1, 2001_11, 2001_12, 2002_8, P11]. Optionally, different laser sources corresponding to different wavelengths (He-Ne-laser at 633 nm, diode laser at 822 nm) can be coupled into the system after passing a Faraday rotator (to avoid parasitic feedback). A chopper (1 kHz) is producing a pulsed signal for *lock-in* measurements. The beam is split at a multichroic beam splitter. One part is detected by a reference detector to measure and compare slight oscillations of the laser source. The other one is focused by an achromatic microscope objective onto the specimen to be characterized. The specimen (e.g. microlens array or diode facet) can be scanned over a maximum transversal path of about 10 mm. The minimum spatial resolution was about 1 μm (depending on objective and wavelength). The reflected beam passes the microscope again and is directed by the beam splitter to the sensitive measuring detector. The detector signals are amplified in two lock-in systems and processed on a computer with a LabView-based analysis software also driving the actuators (step motors). A small part of the beam is temporarily separated by a beam splitter and guided to a CCD camera (for visual evaluation of the adjustment state). The photograph in Fig 98 shows the parts of the experimental setup again (here with 2 laser sources).

The modification of photocurrent and photoluminescence (e.g. for a long-term scale prediction of degradation processes) in diode lasers can also be inspected with the setup.

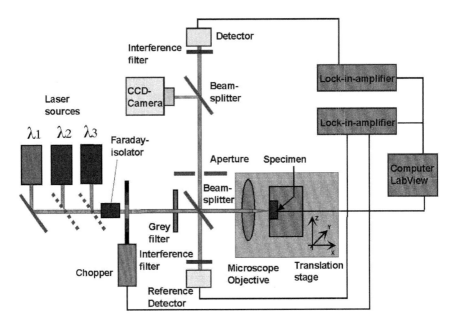

Fig. 97. Setup for the spatially resolved measurement of specular reflectance (schematically). For the explanation, see plain text.

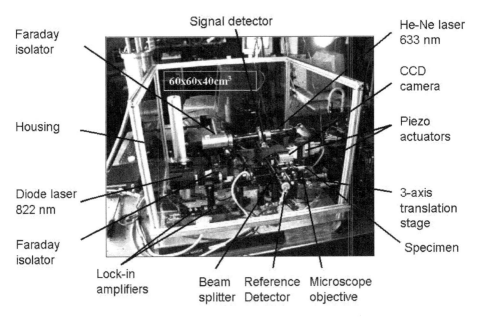

Fig. 98. Experimental setup for the spatially resolved measurement of specular reflectance in lock-in technique.

Furthermore, such measurements can be used to optimize the adjustment of collimating microlenses near diode facets [P10]. The lock-in technique enables to perform *extremely sensitive* measurements at signal-to-noise ratios well below 10^{-4}. This was proved by the inspection of uniform low-reflectance broadband multilayers of known reflectance values (see Fig. 99) as well as the direct detection of the spatial refractive index profile of a low-NA GRIN lens (Fig. 100).

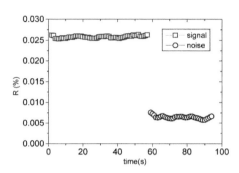

Fig. 99. Experimental prove of the sensitivity of the measuring system: detection of the residual reflectance of a broadband anti-reflection multilayer coating in comparison to the noise level. Signal-to-noise ratios well below 0.0001 were obtained.

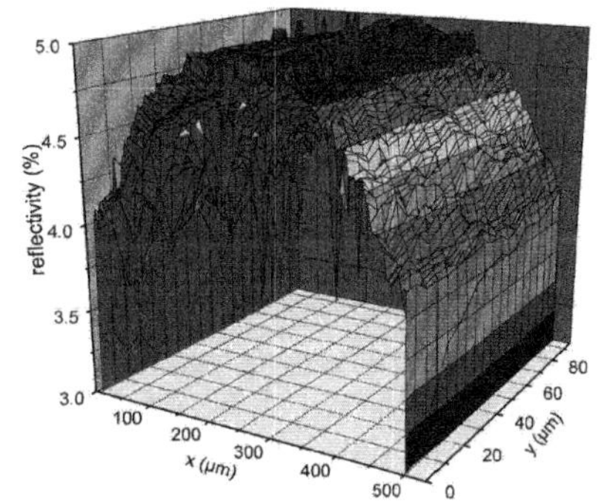

Fig. 100. Spatial reflectivity profile of the facet of a cylindrical low-NA GRIN microlens directly detected by a spatially resolved lock-in reflectance measurement. The total difference in reflectance is < 1%.

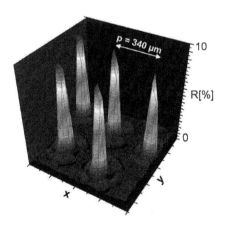

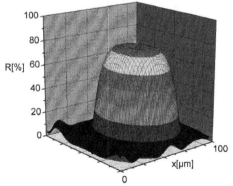

Fig. 101. Reflectance map of a part of a GRMMA of low reflectance (rim zone) at the measuring wavelength (1053 nm). The period of the hexagonal array is 340 μm.

Fig. 102. Reflectance map of a single micro-mirror (mini-GRM) from the center of a GRMMA optimized for a solid-state laser Talbot-resonator (see Chapter 6).

The reflectance profiles of GRMMA consisting of alternating HfO_2 and SiO_2 layers were mapped as shown for a part of a low reflectance array and a single high-reflectance element with super-Gaussian reflectance profile in Figs. 101 and 102. In reflectance mapping, either the measuring wavelength has to be identical with the application wavelength or the real profile has to be calculated on the basis of a reliable multilayer model.

Further applications are the characterization of the uniformity of large-area components, early detection of damage evolution in small areas, detection of biomedical microobjects (bacteria layers), low contrast fingerprints, pick-up of storaged optical data, inspection of fiber endfaces and more.

Polarization aberrations by illuminating the specimen with focused polarized light were theoretically (numerically as well as analytically) studied for the reflection at single layers. It was found that correction factors have to be taken into account under strong focusing conditions, i.e. at large angles of incidence. Figure 103 shows the spatial deformation of the angular spectrum within a focus. The integral error of the reflectance measurement as a function of the divergence angle (inversely related to focal length and spot size) is important for the system calibration because it delivers the necessary *correction terms* (Fig. 104). Further technical improvements (also in spatial resolution) are possible by applying *radially polarized* beams.

Because of the extraordinarily high sensitivity (which was in optimized state comparable with commercial techniques for fiber index measurements with immersion), one of the main targets in developing the system was to find a technical solution for the characterization of graded-AR coatings of small diameter. The results are presented in frame of the applications (see Chapter 6).

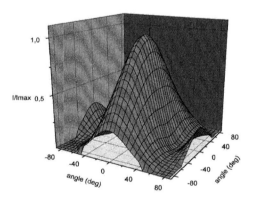

Fig. 103. Polarization aberrations of a Gaussian beam focus induced by the focused illumination with a linearly polarized beam. The normalized intensity distribution as a function of the reflected angle is a measure for the modification of the angular spectrum after the interaction with a plane surface. For the calculation, a total convergence angle of 120° and a refractive index of $n = 1.5$ were assumed.

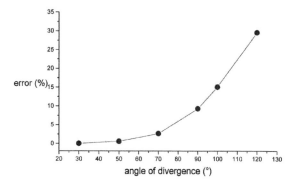

Fig. 104. Polarization aberrations of a Gaussian beam focus with linearly polarized light: integral error as a function of the angle of incidence.

Because of the extraordinarily high sensitivity (which was in optimized state comparable with commercial techniques for fiber index measurements with immersion), one of the main targets in developing the system was to find a technical solution for the characterization of graded-AR coatings of small diameter. The results are presented in frame of the applications (see Chapter 6).

5.3.2. Spatial reflectance mapping with angular resolution

To measure the reflectance distribution *angularly resolved* is a challenging task which was not really solved up to now (cf. sec. 5.1). As a compromise with respect to spatial and angular resolution, a new method based on *truncated or small-angle nondiffracting (Bessel-like) beams* with axially extended focal zones instead of focused Gaussian beams was proposed [2002_16, P11]. The method works at typically by one order of magnitude lower spatial resolution depending on the angular spectrum of the generating components (axicons). More details on this method will be discussed in paragraph 6.1.3. as well.

5.4. Near field propagation measurements

In spatially variable multilayer structures and single-layer thin-film microlenses on substrates of non-identical refractive index, *interference patterns* are generated *within the layers* (e.g. Newton's rings in circular microlenses). If a light beam passes through a very small (few wavelengths diameter) transparent optics of this type, fringe-like intensity distributions can *propagate in the near field* but rapidly will be transformed to the far field pattern because of diffraction. At high intensities, the spatial variations of the near-field intensity, however, strongly influence the efficiency and spatio-temporal uniformity of nonlinear optical processes like second harmonics generation (SHG), multiphoton excitation of molecules, laser damage or image processing. Therefore, a near-field inspection of structured light fields in microoptical systems can be very helpful, especially in the case of complex layer systems. In laser resonators, parasitic near-field patterns could lead to unwanted mode instabilities or hot spots.

The relevance of the problem is illustrated by the fringe propagation of a single photoresist layer on a silica substrate detected with a *scanning near-field optical microscope* (SNOM) in Figs.105a-c. The spatial frequencies near the microlens vertex are very high (up to 3 μm^{-1}).

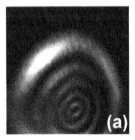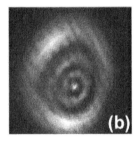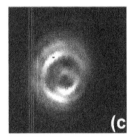

Fig. 105. Near-field propagation of radial substructures generated by internal interference in a small resist layer microlens on a silica substrate (different refractive indices between layer and substrate, lens diameter about 30 μm, 36 μm pitch, layer thickness about 500 nm). The patterns were detected with a SNOM (field of view 12 x 12 μm^2) at different distances from the vertex of the nearly parabolic convex lens: (a) $\Delta z = 20$ μm, (b) $\Delta z = 30$ μm, (c) $\Delta z = 50$ μm. After about 80 μm of propagation, the fringe contrast is almost completely erased by diffraction and refractive focusing (not shown).

Chapter 6

SPATIAL BEAM SHAPING WITH THIN-FILM MICROOPTICS

6.1. Hybrid microoptics for improved efficiency of laser diode collimation

6.1.1. Motivation and basic concepts

Microoptical components are very important for the collimation of high-power laser diodes. To reach high efficiency and mode stability, the following problems have to be solved:

- The beam has to be well collimated.
- The angular distribution has to be well symmetrisized.
- A minimum beam propagation factor (M^2) of the output beam has to be approximated.
- The parasitic feedback (caused by reflections at surfaces of microlenses, fiber end faces etc.) has to be minimized.

In many cases (e.g. for typical diode bars in multi-10-W range), the laser diode radiation is astigmatic and the divergence angles in two different spatial directions (fast axis, slow axis) are extremely different (schematically in Fig. 106). Typically, systems of refractive cylindrical lenses or GRIN structures are used for the beam shaping of such sources [1994_22]. Monolithic systems, i.e. microlenses for fast axis as well as slow axis collimation combined in a single component, are difficult to fabricate and expensive. Sources with better beam properties but lower output power like tapered laser diodes suffer from being extremely sensitive against low-level feedback (schematically in Fig. 107). As it was shown in the literature, a residual reflectance of even 10^{-4} influences the spatial and spectral behavior of the beam significantly [2000_32, pp. 67-81]. Filamentation effects are also know from semiconductor amplifiers [1996_26]. Further problems arise from spatially dependent polarization effects, spectral changes (mode hopping) and beam anisotropy because of 3D smile effects at the diode facets [2002_4, 2002_17, 2004_30]. Therefore, there was a growing motivation to develop:

- new microoptical system design,
- novel types of low-cost, easy-to-fabricate components of high robustness against polarization and small spectral variations.

Two different design concepts for the improvement of the collimation performance were developed on the basis of *thin-film* microoptical components [2001_1, 2004_2, 2004_30, P3, P9] (schematically shown in Fig. 108). The *first approach* [P9] is related to the reduction of the parasitic feed back generated by Fresnel reflection at the plane

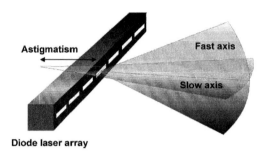

Fig. 106. Astigmatic laser diode array (bar) with significant divergence angles in fast axis and slow axis direction (schematically). The total fast axis angles can reach 90° or more.

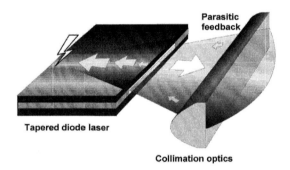

Fig. 107. Parasitic optical feedback into tapered diode lasers by residual reflectance and scattering at optical surfaces (schematically). The reverse amplification of weak signals can lead to instabilities or even destruction of the lasers.

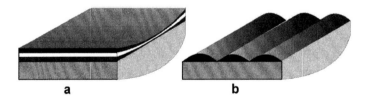

Fig. 108. Two design concepts of hybrid structures incorporating thin-film microlenses: (a) micro-graded AR-coating on a refractive micro-cylinder lens, (b) array of thin-film microlenses for slow-axis collimation deposited on a refractive micro-cylinder lens (FAC-SAC-system) [2004_2].

face of a cylindrical fast-axis-collimation (FAC) lens. It is based on the idea to design a thickness-modulated multilayer as a *graded AR-coating* in such a way that it is well-adapted to the spatio-angular distribution (i.e. the incident angle at each coordinate on the surface) and to integrate it by directly depositing on the lens (schematically in Fig. 109). To realize this approach, *advanced software* for *multilayer design with spatially variable thickness and angle of incidence* and *advanced deposition techniques* with shadow masks had to be developed.

The *second approach* [P3] is based on the *integration of an array of cylindrical thin-film microlenses* for the slow-axis collimation (SAC) in a quasi-monolithic system by directly depositing them onto the plane face of a conventional refractive cylindrical microlens (surface relief or GRIN structure). Thus, a compact and cost-efficient quasi-monolithic FAC-SAC-system can be realized.

Both concepts require improved and even completely new *specific fabrication and measuring techniques*. In particular, the graded AR-design requires inverse Gaussian thickness profiles. These could be obtained by inverting the shadow-mask deposition problem. The two different fabrication schemes for concave and convex layers are illustrated for single elements in Figs. 110a, b.

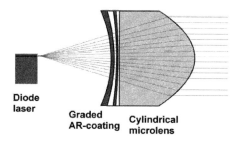

Diode laser

Graded AR-coating **Cylindrical microlens**

Fig. 109. Angular adapted AR-coating designed to minimize the residual reflectance at any spatial coordinate on the surface of a cylindrical FAC microlens (corresponding to Fig. 108a).

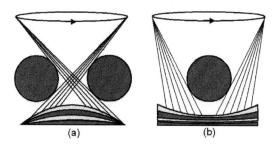

(a)   (b)

Fig. 110. Fabrication of convex (a) and concave (b) layer systems by deposition with cylindrical shadow masks in a rotating system (schematically). The shading configuration in (b) was used to fabricate inverse Gaussian-shaped layer profiles for graded and angular adapted AR-coatings.

For the optimization of FAC-SAC hybrid systems, data on current-dependent near-field wavefront curvature and intensity distributions, polarization profiles, mode behavior and diode facets are needed. For the characterization of graded AR-coatings, high-precision micro-reflectance measurements are necessary. Beside the spatially resolved detection of specular reflectance (as already described in Chapter 5), the spatially resolved measurement of the angular response of multilayer reflectance is of essential interest.

6.1.2. Slow-axis collimation with compact systems of cylindrical microlenses

For the *fabrication* of hybrid FAC-SAC-systems with thin-film microlenses, the techniques of vapor deposition through slit- and wire-grid masks (as reported in detail in Chapter 4) had to be further developed. Fused silica structures had to be deposited on different types of cylindrical microlenses (fiber lenses, rod lenses, GRIN) of varying geometry. High fill factors and appropriate shape functions had to be realized at the same time. The phase profile had to be tolerant against current-dependent variations of the angular structure of the beam. In the result of systematic investigations [2004_30] including the test of many different designs, the specific beam shaping behavior of such FAC-SAC-collimators could be demonstrated experimentally.

Linear arrays of cylindrical SAC-microlenses consisting of fused silica of a maximum layer thickness of 4 µm were deposited with conical slit-array shadow masks

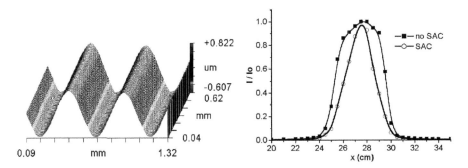

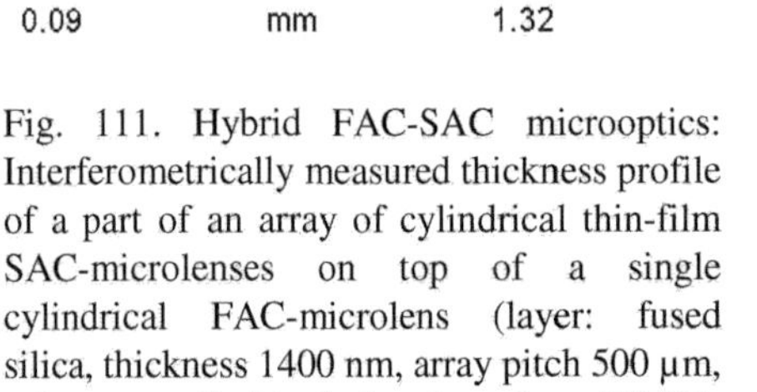

Fig. 111. Hybrid FAC-SAC microoptics: Interferometrically measured thickness profile of a part of an array of cylindrical thin-film SAC-microlenses on top of a single cylindrical FAC-microlens (layer: fused silica, thickness 1400 nm, array pitch 500 µm, substrate: cylindrical glass lens from LIMO). Here, an acceptable compromise between fill factor and shape function was obtained.

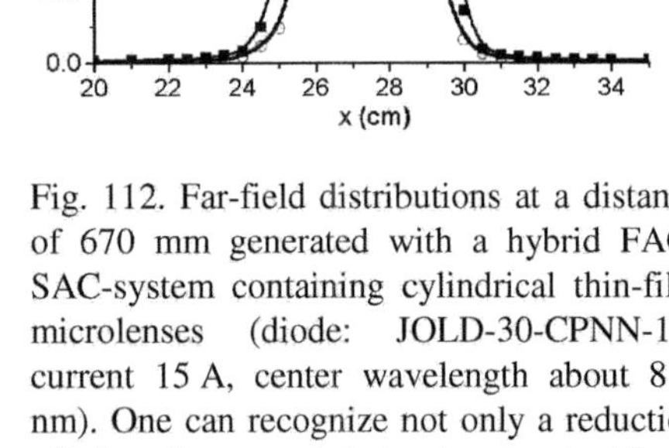

Fig. 112. Far-field distributions at a distance of 670 mm generated with a hybrid FAC-SAC-system containing cylindrical thin-film microlenses (diode: JOLD-30-CPNN-1L, current 15 A, center wavelength about 812 nm). One can recognize not only a reduction of the divergence but also a significant improvement of the distribution (closer to a Gaussian function).

on the flat surfaces of cylindrical FAC-microlenses (LIMO). Fig. 111 shows a part of such an array which was adapted to the geometry of a 30-W laser diode bar. Its fill factor was estimated to be > 94 %. The *collimation performance* was characterized by detecting the far field intensity distribution of a 30-W diode bar (JOLD-30-CPNN-1L, center wavelength about 812 nm) in the slow-axis plane at a distance of 670 mm at currents of up to 25 A (Fig. 112).

As further found in the measurements (not shown in the figure), the field distribution strongly depends on the diode current. A comparison showed a slightly larger dependence on the diode current as a commercial reference SAC microlens array. Our hybrid array reduced the FWHM divergence by a factor up to 0.7 at currents < 15 A, whereas the reference array (with nearly stable divergence) was superior in the region of higher currents (15-25 A.).

The *thickness uniformity* of the elements was characterized interferometrically. The sag height distribution is shown in Fig. 113. The standard deviation was only 0.6%. The rms roughness of the microlenses was excellent as well (about 1 nm, Fig. 114).

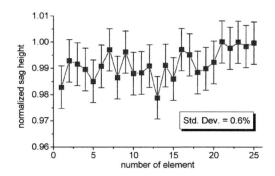

Fig. 113. Characterization of the uniformity of a SAC thin-film microlens array by the interferometrically measured sag height distribution (25 cylindrical lenses, fused silica on glass).

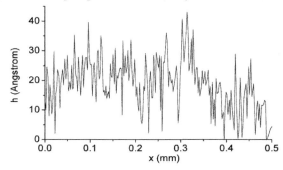

Fig. 114. Roughness of the surface of a fused silica structure on a glass microlens measured interferometrically. The resulting rms value for the cut length of 500 μm was about 1 nm.

From point of view of the fabrication efforts and costs, the hybrid systems with integrated thin-film components have to be considered as an attractive alternative solution. Recent progress in diode laser design promises to simplify the collimation problems in future. The mechanical and thermal robustness could further be improved by transferring the thin-film lenses into bulk glass lenses (e.g. by etching procedures as described in Chapter 4).

6.1.3. Angular-adapted micro-gradient AR coatings

For an optimum design of graded AR-coatings, a perfect adaptation to the incident angular spectrum of a divergent light source is necessary. For extended (e.g. Lambertian) sources, any point on the surface of an optical element represents an angular spectrum instead of a single angle of incidence. The adaptation of a multilayer design has to have a sufficiently high spectral bandwidth in this case. High-power diode sources are not ideal point sources but typically more extended in one spatial direction. Thus it is possible to reduce the problem by assuming a cylindrical angular profile. The minimum multilayer reflectance can be calculated on the basis of a one-dimensional distribution of the incident angles for the half space (because of the symmetry with respect to the optical axis). For the simulations, a new developed software (Multilayer, Kühn Software) was used. In the design calculations, the input thickness profile was chosen to be an inverse Gaussian one. In deposition experiments, this target function had to be well approximated (see Fig. 115). In the region of interest (the central part for the cylindrical FAC-microlens corresponding to the area in Fig. 115), the residual reflectance detected with high spatial resolution was in the range of 0.05 - 0.7 %

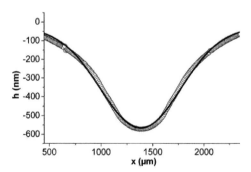

Fig. 115. Thickness profile of a micro-gradient AR-coating deposited on a cylindrical FAC-microlens (interferometrically measured with an auxiliary gold layer to avoid parasitic reflections at the wavelength and angular distribution of the interferometer, magnification 2.5 x). The data points represent the measured profile in FAC-direction (white dots) in comparison to an ideal inverse-Gaussian target function (black line).

(see Fig.116). The considerable advantage of graded AR-coatings in comparison to uniform AR-coatings is demonstrated by the theoretical simulation in Fig. 117.

For the experimental investigation of the *angular dependence* of the residual reflectance, a new technique based on *truncated Bessel-like beams* was used in pilot experiments [2002_16, P 11]. Bessel-like beams show characteristic intensity fringes the envelope function of which is proportional to the square of a Bessel function (see Section 6.3). The depth of focus of a Bessel beam can be chosen to be relatively large (up to > 10 cm). The central peak of such a distribution can be separated by diaphragms (truncation). If the rim of the diaphragm coincides with the first intensity minimum of

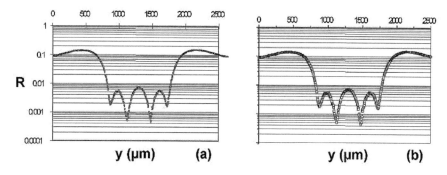

Fig. 116. Spatially resolved reflectance of a graded reflectance AR-coating of inverse-Gaussian (concave) thickness profile: (a) theory, (b) 4-layer system of SiO_2:HfO_2 measured with lock-in micro-reflection technique, see 5.3.1. The outer parts are not relevant with respect to the the lens diameter (800 µm) which is well located within the zone of minimal reflection.

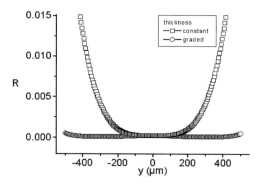

Fig. 117. Theoretical comparison of a graded and a uniform AR-coating with respect to the residual reflectance (optimized case, corresponding to realistic parameters, see Figs. 115, 116). The graded multilayer is obviously stable at values < 0.1 % up to a radius of 400 µm whereas the uniform layer system has already reached a reflectance of 1.5 %. In the experiment, the graded layer had a reflectance of < 0.7 %.

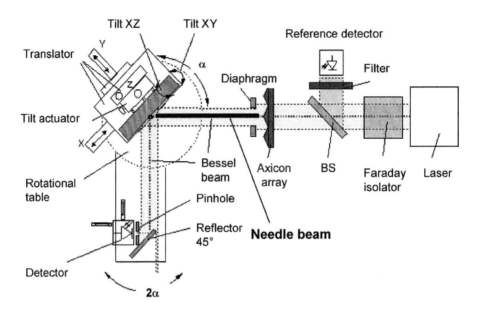

Fig. 118. Setup for angular-resolved and spatially resolved measurement of the specular reflectance of angular-adapted graded multilayer structures. A laser beam is first transformed by an axicon into a Bessel-like beams with concentric intensity fringes. By truncating the Bessel-like beam with a diaphragm of adapted diameter, a single-maximum "needle beam" is shaped which is used to illuminate a specimen at a certain angle α and with a spatial resolution given by its spot size. The mechanical system ensures that the rotation does not shift the illuminated point on the surface (rotational axis fixed). The reflected light is detected with a photodetector and compared to a reference signal similar to the setup described in Chapter 5. (BS = beam splitter.)

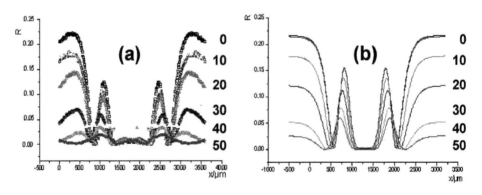

Fig. 119. (a) Measured angular-dependent specular reflectance of micro-graded AR-coatings (a) in comparison to (b) theoretical simulations for angles of incidence between 0 and 50°. Quantitatively, slight differences were observed which are caused by the limitations of both theoretical model and measuring accuracy.

the beam, diffraction effects can be minimized (*self-apodized setup*). The resulting needle-like beam can be used as the measuring beam instead of a Gaussian focus. A setup for the angular- and spatially resolved measurement of the specular reflectance is shown in Fig. 118. The illuminated point on the surface of the specimen stays unchanged if the angle of incidence is changed.

The preliminary results in Figs. 119a and b show a good qualitative agreement between theoretical and experimental curves but some quantitative deviations which are caused by the accuracy of simulation and measurement. Both have to be further developed yet. The use of truncated Bessel-like beams for spatially and angularly resolved measurements is also of interest for *other applications* like ellipsometry.

6.2. Mode-selection in laser resonators

6.2.1. Stability management of compact solid-state laser resonators with micro-mirrors

The spatial beam-shaping capabilities of thin-film microoptical components can also be used to *select and to stabilize laser modes* [1999_8]. This was demonstrated by different types of SiO_2:HfO_2 - multilayer structures as output coupling mirrors in compact, diode-pumped Nd-doped solid-state lasers (setup in Fig. 120). In Fig. 121, three variants of mode-selective mirrors are shown.

By comparing the *stability behavior* to the case of a macroscopic mirror it was demonstrated that a similar operation characteristics (at a reduced output power) can be achieved with a miniaturized laser of much shorter resonator length, Fig. 122.

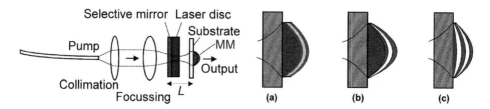

Fig. 120. Setup of a compact diode-pumped solid-state laser system with a thin-film micro-mirror (MM) as a mode-selective output coupler. The selective dichroic mirror is highly transmitting for the pump wavelength but highly reflecting for the laser wavelength. (L – resonator length.)

Fig. 121. Selected types of multilayer and compound mirrors for mode selection in resonators: (a) dielectric microlens coated with *uniform* multilayer, (b) dielectric *microlens coated* with *nonuniform* multilayer (graded reflectance micro-mirror, GRMM), (c) nonuniform multilayer GRMM.

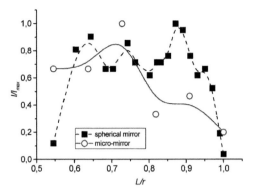

Fig. 122. Tailoring the resonator stability of a micro-disk solid state laser by inserting a graded reflectance micro-mirror (type as in Fig. 121c). The laser performance (normalized intensity) with micro-mirror (circles, straight line) is compared to a larger setup with a spherical mirror (squares, dotted line) for the same values of L/r (resonator length / radius of curvature).

6.2.2. Talbot resonators with micro-mirror arrays

Talbot resonators exploit the self-imaging properties of periodic arrangements of phase and/or amplitude objects (see Chapter 2). If self-imaging is generated within a resonator structure (e.g. by micro-mirror arrays) and the resonator lengths is chosen to be a multiple of the Talbot distance, a resonant self-imaging can be obtained where the original phase and/or amplitude information is replicated directly at the mirrors. Coupled (multiple) self-imaging resonators can also be constructed [P2]. The mode volume of Talbot resonators can be optimized by varying the shape of the mirror reflectance curves and envelope phase functions [1994_2]. The modification of self-imaging patterns by envelope functions is shown in Fig. 123 [1998_17, p. 88].

Self-imaging in Talbot cavities has first been applied to the phase-locking of laser diode arrays [1989_21, 1990_11, 1994_2]. Spatial loss modulation with intra-cavity wire-grids satisfying the Talbot imaging condition has been used to stabilize CO_2-lasers in single high-order lateral mode regime [1992_17] as well as to phase-lock arrays of such lasers. Talbot modes have been studied theoretically for a solid-state laser arrangement longitudinally pumped by an array of 5 x 5 laser diodes [1993_24]. As an alternative solution where the high internal losses at absorbing spatial filters are avoided, the use of modulated reflectance has been suggested [1989_20]. With graded reflectance micro-mirror arrays (GRMMA) it was possible to realize such a particular *self-imaging laser architecture* for the first time [1993_5, 1994_2].

A flash-lamp pumped high-power Nd:glass laser with Talbot resonator was built up to demonstrate the generation of beam arrays [1993_5, 1994_2, 1995_3]. The laser consisted of a 9 cm long Nd:glass rod of 3 mm diameter (endfaces tapered to 4 mm).

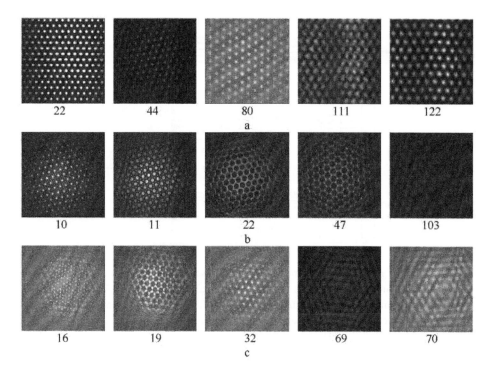

Fig. 123. Self-imaging of GRMMA with and without envelope phase functions: Intensity distribution detected with a CCD camera at different distances from the array, (a) hard edges (higher-order super-Gaussian reflectance profile), (b) soft edges (Gaussian reflectance profile), (c) hard edges and envelope Gaussian thickness function (GRMMA: alternating SiO_2 and HfO_2 layers on BK7, period 340 µm, wavelength 1054 nm, field of view: 4.9 x 4.9 mm^2).

The resonator scheme is drawn schematically in Fig. 124. Different types of hexagonal GRMMA (hard edges, soft edges and soft edges with envelope functions, up to 1250 elements of 340 µm pitch) were applied. The maximum center reflectance of the hard and soft elements was about 55 and 30%, respectively. the GRMMA consisted of 27 alternating dielectric layers (14 HfO_2, 13 SiO_2) of profile heights of 1.8 µm (soft mirror) and 2.3 µm (hard mirror). Position and angular alignment of the mirrors were adjusted by monitoring the intensity patterns at the GRMMA with a CCD camera. In the case of the envelope reflectance function, this was much easier because the micro-mirror of maximum reflectance in the center can be used as reference point. The optimum total length of the resonator in air was found to be L = 35 cm (corresponding to an optical path of 41 cm and the 4th Talbot plane for 1 roundtrip).

Matrix-shaped beams of up to 200 phase-coupled partial beams of a total output energy of up to 425 mJ were generated at a pump energy of 790 J and a pulse repetition frequency of 1 Hz [1995_3]. The far field has been measured in the focus of a lens with

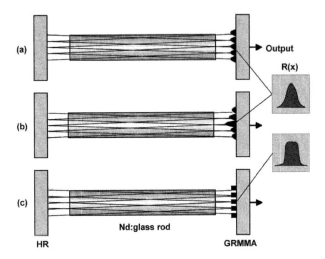

Fig. 124. Talbot resonator configurations for Nd:glass laser containing different types of GRMMA as output coupling mirrors (schematically): (a) uniform GRMMA with soft mirrors (Gaussian reflectance profile), (b) nonuniform GRMMA with soft mirrors and macroscopic Gaussian envelope thickness function, (c) uniform array with hard-edged mirrors (super-Gaussian reflectance profile). The types of reflectance profiles are sketched in the insets (right).

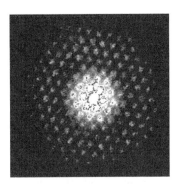

Fig. 125. Far field of a beam array of about 200 partial beams generated with a flash-lamp pumped Nd:glass laser with a Talbot resonator containing GRMMA as output coupling mirrors (detector: LBA 100 Spiricon system). The angular difference between two adjacent beams is 3.2 mrad. Most of the energy is concentrated in the 7 central spots (picture overexposed to show also the outer parts).

$f = 1$ m (Fig. 125). The divergence angles of the single beams and the envelope of the beam array were determined to be 2.3 mrad and 16 mrad, respectively.

Because of their phase profiles, soft (approximately Gaussian-shaped) GRM act like microlenses so that the beam array is *focused* without additional lenses. The focal lengths of the GRMMA in transmission at a wavelength of 1.06 μm was f = 15 mm. The intensity envelope of the beam array in the near field had a $1/e^2$-diameter of 1.5 mm. The high intra-cavity power density was used for passive Q-switching in Cr:YAG crystals yielding 25 ns pulses at pulse energies of 250 mJ (at a pump energy of 830 J).

6.2.3. Mode stabilization in laser diode MOPA-systems with external resonator for second harmonic generation

UV emission from cw laser diode based systems via direct second harmonic generation (SHG) is difficult because the conversion efficiency depends on the square of the intensity. To enhance the efficiency, laser diode master oscillator and power amplifier systems (MOPA) plus resonance enhancement in external resonators are used [1999_9]. Fig 126 shows a setup with a selective feedback loop which couples a small amount of light energy back to the oscillator diode to stabilize the frequency.

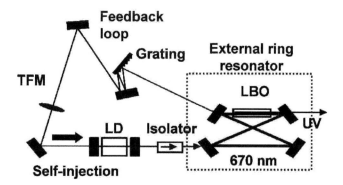

Fig. 126. Diode laser system with external bow-tie resonator for resonance enhanced SHG in LBO and a grating-tunable feedback loop (LD - laser diode MOPA consisting of an oscillator diode and a broad area diode amplifier, LBO - Lithium Borate crystal for SHG, TFM - thin-film microlens). The diode laser wavelength of 670 nm is directly converted to a UV wavelength of 335 nm. The thin-film microlens TFM has a Gaussian-shaped profile (micro-axicon) and acts as directional filter which stabilized the feedback in the spatial domain.

In the experiment, a 7.8 mW oscillator diode (670 nm) was isolated by a double-stage Faraday rotator and amplified in a broad area diode amplifier. After mode filtering, about 80 mW were inserted into an external resonator of high Q-factor (total losses 1.6% depending on the conversion process). Enhancement factors > 60 and a circulating power of 5 W in the ring resonator were obtained. The UV output power at 335 nm was typically between 0.3 and 1 mW (in pulsed mode up to 100 mW). For a spectral fine tuning of the feedback, a grating was used. To stabilize the laser performance, the Hänsch-Couillaud method [1980_4] was applied. However, a perfect stabilization was not obtained because of the counteracting time constants and the large number of influence factors (long optical paths, limitations in the pointing stability of the laser diodes, thermal effects, vibrations, air fluctuations etc.) which make such systems at least extremely sensitive against small perturbations.

It could be demonstrated that the implementation of a single *micro-axicon lens* of Gaussian phase profiles as a *directional filter* improved drastically the stability (TFM in Fig. 126). With the directional filter, stable operation was observed over many hours. This is illustrated by the temporal evolution of the mode *spectrum* without and with thin-film microlens filter (Figs. 127a,b) and corresponding UV mode intensity patterns in *space* (Figs. 128a,b). The spatial stability was documented in a video record over about 1 h. The direct connection between spectral and spatial dynamics was verified by a spectrally and temporally resolved multichannel measurement with an array of fast photodiodes (time resolution about 0.5 ns).

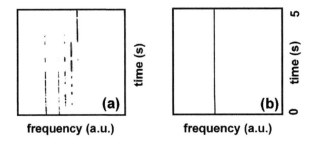

Fig. 127. Temporal evolution of the UV-modes in the frequency domain: (a) without directional filter, (b) with thin-film microlens (TFM) as directional filter (detection system: SOPRA spectrometer and Coherent linear CCD array, exposure time 17.41 ms, pixel frequency 42 Hz).

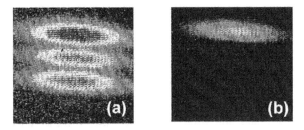

Fig. 128. Mode pattern of the UV-modes in spatial domain: (a) without directional filter, (b) with thin-film microlens (TFM) as directional filter (detection with UV-converter and CCD-camera).

6.3. Generation of Bessel-like nondiffracting beams

6.3.1. Nondiffracting beams

As it was demonstrated by simulations and experiments (see Chapter 4), spherical as well as *nonspherical thickness profiles* of thin film microoptical components can be realized by shadow mask vapor deposition techniques. In particular, Gaussian and inverse-Gaussian shaped structures and double-parabolic profiles can easily be fabricated. The characteristic angular distributions which can be generated with such refractive or reflective devices enable to approximate so-called *nondiffracting beams* or (physically more correct) *pseudo-nondiffracting beams*.

Historically, the idea of "*nondiffracting*", "*diffraction-free*", "*limited diffraction*" or "*non-spreading*" light waves and "*focus wave modes*" was born several times at different places and in different context. In particular, it was developed in the frame of the theory of Maxwell's equations, optical engineering, refractive laser beam shaping, diffractive optics and holography. In 1941, Stratton predicted so-called "undistorted progressive waves" [1941_1]. Mc Leod solved the problem of the limited depth of focus of slide projectors by inventing a new class of optical components which he referred to as *AXICONS* (i.e. elements shaping a CONstant focus along the optical AXIs). [1954_1]. Axicons are of nonspherical shape (but not limited to perfect cones, as many opticians believe - he proposed various types in his original paper!). The capability to generate *extended focal zones* along the optical axis [1986_5, 1989_7] is of course highly interesting for many other applications as well. Recently, for example, propagation-invariant zones of > 20 cm were realized with lithographically written conical microaxicons of 1mm diameter for applications in optical particle manipulation [2005_4].

In theoretical studies it was shown that *propagation-invariant solutions of the Helmholtz wave equation* can be found under certain boundary conditions (see the papers of Durnin [1987_8, 1987_9], Brittingham [1983_3], Ziolkowski [1985_6, 1989_6, 1989_7], Indebetouw [1989_5]) which can be extended to generalized classes of nondiffracting wave phenomena [1996_10, 1998_10, 2000_16, 2002_11, 2004_20] with light, THz radiation, radiowaves as well as acoustic waves including vector beams [1991_8, 2002_45, 2005_32]. Durnin demonstrated that the superposition of convergent partial beams (e.g. in the Fourier plane of a annular slit) leads to the stable axial propagation of fringe-like transverse intensity patterns by *repeated constructive interference* all along the axis. The field distribution $E(r,t)$ in radially symmetric direction case can be described by a Bessel function J_0 of zero[th] order and first kind so that this type of beams is called *Bessel beams* [1987_8]:

$$E(r,t) = \exp[i \cdot (k_z - \omega \cdot t] \cdot J_0[k_r \cdot r] \qquad (104)$$

with
$$k_r = \frac{2\pi}{\lambda}\sin\beta \tag{105}$$

and
$$k_z = \frac{2\pi}{\lambda}\cos\beta \tag{106}$$

(k_r, k_z - radial and axial components of the wave vector, r - radial coordinate, β - conical beam angle, $\omega = 2\pi\nu$ is the angular frequency related to the frequency ν). The conical beam angle is the angle of the partial beams against the optical axis. The intensity profile is governed by the square of the Bessel function (Fig. 129). The propagation of Bessel beams is significantly different to beams of a Gaussian profile [1988_1]. *Ideal Bessel beams* are generated by transversally unlimited apertures and therefore propagate over *infinite distances* carrying *infinite energy*. The energy of all fringes is then identical but the highest intensity is found in the central lobe.

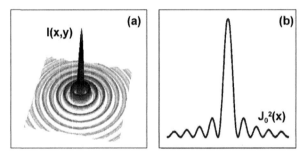

Fig. 129. Intensity distribution of an ideal Bessel beam proportional to the square of the zero[th] order Bessel function: (a) intensity as a function of both spatial coordinates x,y with the coordinate origin in the center; (b) Square of the zero[th] order Bessel function in one spatial direction (calculated). The minima correspond to zero intensity.

Because of the confusions typically caused by the term "nondiffracting", it should be pointed out that in every case the propagation means a propagation of a resulting interference pattern (the propagation of which has clearly to be distinguished from the propagation of constituting conical beams coupled to their Poynting vector [2002_20]).

Bessel beams and related nondiffracting beams can show some unusual properties like considerable *axial field components, Gouy phase shift* (known from any focused electromagnetic waves [1890_1, 1890_2, 1980_3, 1999_23, 2006_29]) combined with *superluminal* group velocities (without violating the basic laws of physics) [1997_42, 1999_13, 1996_7, 2001_18], or *spatial self-reconstruction* in linear or nonlinear media [1997_41, 1998_11, 2001_35, 2002_45, 2006_5]. The Gouy phase shift can be interpreted as a geometrical quantum effect arising from the uncertainty principle in transverse spatial confinement [1996_29, 2001_36]. The self-reconstruction of Bessel beams follows logically from the principle of their generation. It is combined with transverse-to-axial transfer of spatially encoded information: Any shading object of finite diameter placed in a conical beam at a finite distance produces a *shadow zone* of finite axial extension. Behind this zone, the interference pattern can be generated anew without disturbances.

All of these properties make Bessel beams very interesting for *laser applications* like optical acceleration [1991_26, 1997_43, 2000_35], nonlinear optics (harmonics generation [1993_25, 1999_11, 2000_34] including SHG in lossy media [2002_46], Čerenkov phase matching [1997_45], filament control, stimulated Raman scattering [1996_30], Z-scan method [1997_44], multiphoton ionization, photorefractive two-wave mixing [2003_31]), temporal shaping femtosecond pulses [1996_7, 1999_24], particle manipulation, materials processing [1996_6], or laser triggered switching of electrical discharges. Laser resonators with axicon mirrors (*Bessel resonators*) were developed [1989_8, 2001_15, 2005_16]. Because of their high applicability, the propagation of Bessel-beams through axi-symmetric optical systems [1991_6] and the performance of complex nondiffracting beams and arrays [2002_26] were studied systematically.

6.3.2. Bessel-like beams

In *real world experiments*, nondiffracting beams can only by approximated by *pseudo-nondiffracting beams* of finite axial extension and finite energy (referred to as *Bessel-like beams* by Herman and Wiggins [1991_5]). If the interference pattern is shaped by conical lenses or spherical lenses with significant spherical aberrations, the stationary field distribution can be approximated by a zero[th] order Bessel function J_0:

$$E(r) \propto J_0[k \cdot r \cdot \sin(\beta)] \tag{107}$$

($k = 2\pi/\lambda$, r - radial coordinate, β - conical beam angle). The scheme in Fig. 130 shows the geometrical situation for the generation of Bessel-like interference zones with a single radially-symmetric refractive axicon.

In the more general case, the angular spectrum has not necessarily to be narrow so that eq. (107) can slightly be modified by an *axial dependence of the angle of incidence* $\beta(z)$ (z - axial coordinate, Figs. 131, 132):

$$E(r) \propto J_0\{k \cdot r \cdot \sin[\beta(z)]\} \tag{108}$$

The maximum depth of the focus or "Bessel zone" (i.e. the region of interference of the constituting partial beams) is limited by the diameter of the generating field in the start aperture and the angular spectrum of the initial wave [1993_9, 2000_36].

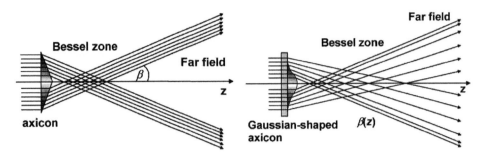

Fig. 130. Approximation of nondiffracting beams by cone-shaped refractive axicons (schematically). Here, Bessel-zones of finite extension are shaped. The conical beam angle (angle of refracted beams against the optical axis) is β. One can easily recognize that the narrow angular spectrum leads to a ring shaped far field distribution which is characteristic for Bessel beams. The transversal spatial frequency in the Bessel zone is constant (if diffraction effects are neglected).

Fig. 131. Approximation of nondiffracting beams by Gaussian-shaped refractive axicons (schematically). Compared to the case of cone-shaped axicons in Fig. 130, extended Bessel-zones are obtained. The conical beam angle β is now a function of the axial position z. The wide angular spectrum in this case leads to a significant broadening of the far field ring pattern. The transversal spatial frequency in the Bessel zone is slightly varying with the axial position.

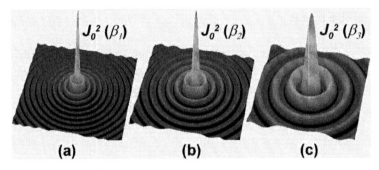

Fig. 132. Variation of the spatial frequency of the Bessel-like fringe patterns with the propagation-dependent angle of incidence (schematically): (a) highest spatial frequency corresponding to an incident angle of β_1, (b) medium spatial frequency corresponding to β_2, (c) low spatial frequency corresponding to β_3, with the relations $\beta_1 > \beta_2 > \beta_3$.

Additional modifications are introduced by diffraction at the aperture which generates a separate angular spectrum and has to be taken into account at low Fresnel numbers (large distances).

A special type of pseudo-nondiffracting beams of large practical impact are the so-called *Bessel-Gauss beams* (BGB) [1987_2, 2002_19]. BGB start from a Gaussian beam instead of a plane wave what is very typical for experimental setups with laser sources. The beam propagation factor for BGB is different from that for Gaussian beams [1997_13].

For optical tweezers, atom guiding and related applications, the focusing of *Bessel-Laguerre beams* by axicons is important because of the central minima in the intensity distribution and their topological charge (vortex in the center and orbital momentum) [1999_14, 2000_17, 2002_13, 2002_15, 2002_44, 2003_14, 2005_18]. In general, the implementation of *amplitude and/or phase masks* in the initial wave leads to *modified Bessel-like beams* [1994_5]. Beam shaping with *continuously varying conical angles* was studied [2003_15]. It was shown that additional degrees of freedom arise from the *interference* of multiple nondiffracting beams [1998_7].

A further important type of nondiffracting beams are the so-called *Mathieu beams* [2000_20, 2000_21, 2001_20, 2005_18] which are described by Mathieu functions and can be extended to helical [2006_12] and high-order [2002_22] subspecies. Mathieu beams appear to be degenerated Bessel beams with distinct fringes in only one dimension and can be approximated by axicon mirrors under off-axis conditions. Their generalization is the class of *elliptical nondiffracting beams* [2002_21].

Pseudo-Bessel beams were realized by different types of optical elements or systems which create the necessary conical distribution (e.g. refractive axicons, refractive lenses [1994_5] or lens systems [1987_3, 1991_10, 1993_8, 1989_9], holographic DOE [1991_7, 1988_2], conical mirrors [2002_18] and programmable spatial light modulators [1996_31, 1996_32, 2006_31]. Recently, the generation of diffraction-free near-field patterns was proposed [2004_17]. Other current topics are the influence of partial coherence [2005_13] and the propagation of polychromatic nondiffracting beams [2005_20] as produced by femtosecond lasers.

6.3.3. Generation of arrays of microscopic Bessel-like beams with thin-film axicons

It was demonstrated by the group of the author that non-spherical *thin-film structures (thin-film micro-axicons)* can also be used to generate Bessel-like beams of extended as well as oscillating or multiple focus zones [1997_8, 1997_10, 1998_3, 2000_4, 2001_7, 2001_8, 2001_9, 2002_5]. In the first experiments, thin dielectric layers consisting of materials like fused silica, HfO_2 or TiO_2 were deposited on silica and glass substrates. With shadow mask arrays, hexagonal or orthogonal *arrays* of circular microaxicon lenses as well as linear arrays of cylindrical elements of Gaussian cross-section were obtained (for the details, see Chapter 4). With such axicons, all the specific advantages of thin-film microoptical arrays as discussed in the previous Chapters can be combined with the particular features of pseudo-nondiffracting beams [2000_4, 2000_6, 2004_2, 2004_9, 2004_10, 2005_5, 2005_6]. First of all, these are:

- generation of small conical beam angles (for very extended focal zones; for the generation of low and experimentally resolvable spatial frequencies; for the reduction of spatial dispersion effects) [2005_06, 2005_10]
- high tilt tolerance (for strongly curved wavefronts, e.g. in wavefront sensors; for reflective setups) [2002_6, 2003_1, 2004_14, 2004_19, 2006_7]
- low dispersion and broadband spectral transfer (for ultrashort-pulse applications [2004_15])
- array design (for optical matrix processors [2003_11, 2004_2] and multichannel materials processing)
- VUV-capability (for excimer laser and Ti:sapphire laser harmonics beam shaping) [2004_1, 2005_9, 2006_6]

Before the experimental realization of these points and specific applications will be addressed in the following paragraphs, essential beam shaping conditions for *refractive thin-film Gaussian-shaped microaxicons* will briefly be analyzed.

The first remarkable difference to any other type of axicons is that with *ultraflat* elements pseudo-nondiffracting beams of unprecedentedly small conical beam angles can physically be realized. For beam shapers of arrayed microaxicons, where the single elements are small compared to the illuminating beam size, in many cases the illumination for each element can approximately be regarded as a *plane wave*. This is a second difference to setups including macroscopic axicons. A further specialty is the Gaussian shaped phase generating a particular type of pseudo-nondiffracting beams. The spatio-temporal characteristics of pulsed beams of this type was analyzed in detail by Kebbel [2004_21].

Assuming axially symmetric geometry and on-axis axicon design, the angular behavior can be described by a simple model. For practical reasons, the orientation of the microaxicons in propagation direction is favored as shown in the schematic drawing in Fig. 133. Only in this case, the minimal propagation distance can be used for experiments (otherwise, a part of the propagation path is located inside the substrate and may cause nonlinear effects at high intensities). Furthermore, only one refractive surface has to be taken into account for simulating the refractive beam shaping. A plane-parallel substrate is then passed by a parallel monochromatic beam without any wavefront distortions. A light ray at a distance r from the optical axis arrives at the refracting layer surface (from inside the layer) at an angle of α to the normal on the surface point. After being refracted it leaves the axicon at an angle β to the optical axis. The *angular distribution* $\beta(r)$ caused by a refractive thin-film axicon of a refractive index n and a radial thickness profiles $h(r)$ is given by [2001_7]

$$\beta(r) = \arcsin(n \cdot \sin\alpha) - \alpha(r) \qquad (109)$$

where

$$\alpha(r) = \arctan\left[\frac{d}{dr}h(r)\right]. \qquad (110)$$

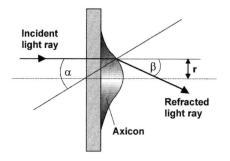

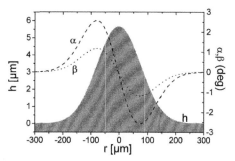

Fig. 133. Angular transformation of a parallel beam at a Gaussian-shaped refractive thin-film microaxicon (α incident beam, β refracted beam, schematically).

Fig. 134. Thickness profile $h(r)$ (gray) and angular profiles $\alpha(r)$ (dashed line) and β (r) (dotted line) for single thin-film microaxicon (fused silica deposited on silica, vertex height $h_{max} = 5.64$ µm, refractive index $n = 1.46$ at the reference wavelength of $\lambda = 790$ nm).

For a Gaussian-shaped axicon with the profile parameters h_{max} (vertex height) and w_0 (waist), one obtains the angular distribution of the refracted beams as follows:

$$\alpha_G(r) = \arctan\left[-\frac{4}{w_0^2}rh_{max}\exp\left(-2\frac{r^2}{w_0^2}\right)\right] . \tag{111}$$

Thickness profile and angular distribution of a single axicon of 5.64 µm vertex height with a refractive index of $n = 1.46$ at a wavelength of 790 nm are shown in Fig. 134. The corresponding spatial frequencies of the Bessel-like fringes in transverse direction follow

$$\nu_S(\beta) = \frac{2\sin\beta}{\lambda} = \frac{1}{\Lambda_S} \tag{112}$$

(Λ_S - fringe distance, β - refracted angle, λ - wavelength). In the experiments, *arrays* of up to $> 10^3$ microscopic-size Bessel-like beams were generated by using low-dispersion refractive, reflective and pseudo-reflective structures as beam shapers (design and fabrication see Chapter 4). In Fig. 135 such an array of Bessel-like beams is demonstrated by a cut through the intensity distribution within a plane behind the array of axicons corresponding to Fig. 134. The illuminating beam stems from a 10-fs Ti:sapphire laser oscillator. The intensity pattern was imaged onto a CCD camera with a microscope objective and a zoom lens. For ultrashort-pulse applications of refractive axicons, the dispersion of the substrate (0.5-1 mm) was pre-compensated. The effective number of partial pseudo-nondiffracting beams varies with the diameter of the illuminating laser beam.

By varying the parameters of refined types of thin-film microaxicons and systematically exploiting their specific advantages it became possible to tailor beam

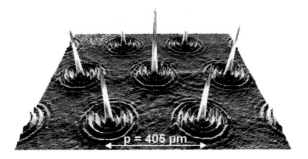

Fig. 135. Transverse intensity profile of a part of a hexagonal array of microscopic-size Bessel-like beams detected in the plane of maximum fringe contrast and maximum axial stability of the fringe pattern (axial distance of the imaged plane: z = 9 mm, array period: 405 µm).

arrays as light tools for new and exciting optical applications, in particular for the localization of light in needle-shaped propagation volumina, multichannel processing of optical data and materials under extreme conditions, ultrafast optical measuring techniques and the sophisticated characterization of very shortest wavepackets in space and time.

6.3.4. Spatial self-reconstruction of Bessel-like beams

Recently, the spatial self-reconstruction properties of single and array-shaped nondiffracting beams were demonstrated with thin-film microaxicons for the first time

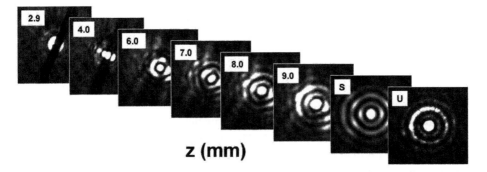

Fig. 136. Spatial self-reconstruction of a single microscopic-size pulsed Bessel-like beam. A Ti:sapphire laser beam (12-fs pulse duration, center wavelength 790 nm) passed a thin-film microaxicon (fused silica on silica, Gaussian surface shape, parameters see Fig. 134) and was distorted by a wire at z = 2.9 mm behind the axicon. The pattern of the propagating beam was imaged with a CCD-camera at the notified distances z. Picture "U" represents the undistorted pattern at z = 9 mm, "S" the corresponding computer simulation. Field of view is 217 x 217 µm².

[2004_12, 2006_5]. Figure 136 shows the evolution of a single Bessel-like wave field generated from a Ti:sapphire laser beam (oscillator, pulse duration about 12 fs) after significantly disturbing the propagation by a reflecting cylindrical object (gold wire, 30 µm diameter, distance 2.9 mm from axicon vertex). The results in the pictures are compared to the undistorted case and a corresponding simulation. One can clearly recognize the *recovery of the fringe pattern* after leaving the zone of distortion. The continuously increasing contrast with the distance allows for the identification of *specific information* (spatial frequency, Bessel fringe structure) even at an early stage of reconstruction so that Bessel beams should be very promising candidates for optical data transfer.

Further self-reconstruction experiments were performed in spatio-temporal domain and with microaxicon arrays combined with spatial light modulators.

6.3.5. Self-apodized truncation of Bessel beams as a first way to shape needle beams

As explained in paragraphs 6.3.1. and 6.3.2., the transverse intensity pattern of Bessel-like beams consists of characteristic concentric *fringes*. However, if the application requires a focal zone (*spatial localization*) of only a *single intensity maximum* but axially extended, and a reduced total energy is acceptable, the truncation of a Bessel-like beam by means of a diaphragm can be a solution of the problem [1995_4, 1997_12, P11]. The truncation of the fringes with masks of different diameter is schematically depicted in Fig. 137 (above). A single-maximum beam is obtained if only the central maximum can pass the aperture. To avoid diffraction at the rims of the diaphragm, the edge of the aperture has to be matched to the minima of the electrical field [2004_11, 2005_8, 2005_12]. This configuration can be referred to as "*self-apodized*" setup [2004_11] because it takes advantage of the smooth transversal amplitude profile (central lobe of the Bessel function) itself instead of inserting an apodizing amplitude mask. This situation is sketched in the lower part of Fig. 137.

The shaping of intensity distributions similar to self-apodized Bessel functions is well known from guided modes in gas-filled *hollow waveguides* for efficient high harmonics generation with intense femtosecond laser pulses [2002_47, 2003_32].

Because of their large aspect ratio (i.e. the ratio between axial and transverse extension), truncated Bessel-like beams have also been called "needle beams" (not to be confused with the *Nadelstrahlung* [1917_1] which Einstein introduced as a model for the directionality and momentum of single photons). The spatial dimensions of truncated Bessel zones can be illustrated by simple geometrical considerations. In the case of an ideal Bessel beam, the diameter of the central maximum d_0 can easily be calculated by the first zero of the Bessel function:

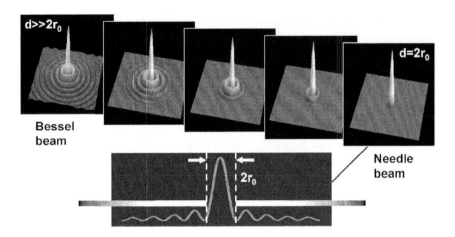

Fig. 137. Generation of single-maximum beams or "needle beams" by self-apodized truncation of Bessel-like beams (schematically). The number of fringes passing an aperture of decreasing diameter (from left to right, above) is decreasing as well. If the diameter of the aperture is small enough, only a single maximum is obtained. To avoid diffraction, the edge of the aperture has to coincide with the first minimum of the Bessel distribution ("self-apodized setup").

$$d_0 = 2r_0 = \frac{3\lambda}{4 \cdot \sin\beta} \qquad (113)$$

where β is the conical beam angle and λ the wavelength. In a simple geometrical-optical picture, the axial extension of a Bessel zone Δz_B can be estimated from the radius of the illuminated part of the axicon r_{max} to be

$$\Delta z_B = \frac{r_{max}}{\tan\beta} \ . \qquad (114)$$

In the case of self-apodized truncation ($r_0 = r_{max}$), the Bessel zone reduces to:

$$\Delta z_{BT} = \frac{r_0}{\tan\beta} \ . \qquad (115)$$

With the help of the trigonometric relation [1995_30, p. 310]

$$\arctan(x) = \arcsin\left(\frac{x}{\sqrt{1+x^2}}\right) \qquad (116)$$

and taking into account that all arguments in the trigonometric functions in eqs. (113 and 114) are positive, Δz_{BT} can also be expressed as a function of the diameter of the central maximum:

$$\Delta z_{BT} = \frac{4d_0^2 \cdot \sqrt{1-\left(\frac{3\lambda}{4d_0}\right)^2}}{3\lambda} . \qquad (117)$$

For $\lambda \ll d_0$, one finds the approximation:

$$\Delta z_{BT} \approx \frac{4d_0^2}{3\lambda} . \qquad (118)$$

This relation can be interpreted as a *propagation law* for the axial and radial extensions Δz and Δd of Bessel zones of arbitrary radially symmetric truncation and is proportional to the reciprocal Fresnel number of the system:

$$\frac{\Delta z}{\Delta d^2} \cdot \lambda \approx const . \qquad (119)$$

Under real conditions with non-perfect pseudo-nondiffracting beams in the presence of diffraction, refractive cross-talk and broad spectra of complex structure, one has to determine the optimum diameter of the truncating aperture experimentally. The diaphragm should consist of a dielectric material to avoid its interaction with the magnetic field component which reaches its maximum value at just the zero intensity positions of the electrical field component [2005_37]. If the application is phase sensitive one should take into account that slight phase corrections and a focal shift are induced by the truncation [2000_11, 1996_33, 1997_12].

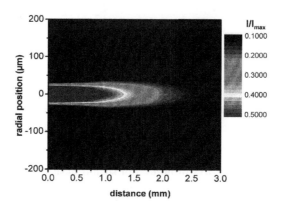

Fig. 138. Truncated microscopic-size Bessel-like beam generated from a 10-fs Ti:sapphire laser beam by a thin-film microaxicon of Gaussian shape (thickness about 6 μm). A single beam was separated from an array of beams (hexagonal arrangement, period 405 μm) by self-apodized truncation with a pinhole of about 30 μm diameter placed in the center of the Bessel zone. Please notice the different scales. (In a 1:1 presentation, the needle-like shape of the beam would lead to a vertical shrinking of the distribution to a thin horizontal line).

In Fig. 138, the propagation of the radial intensity profile of a pulsed single-maximum Bessel-like beam ("needle beam") after self-apodized truncation is plotted. The intensity patterns were detected by imaging the planes of interest onto a 4-million-pixel CCD camera. For the truncation, a 30-µm aperture was placed in the center of the Bessel zone at $z = 9$ mm (highest intensity and contrast). The axicon parameters correspond to the data from Fig. 134. The propagation is obviously almost not distorted by the diaphragm (no diffraction ripples visible) despite the broad bandwidth of the ultrashort pulse (FWHM about 120 nm). The Rayleigh length of the truncated Bessel zone was found to be about 1 mm. The aspect ratio (ratio of the radius of the aperture to the Rayleigh length) was 130:1. With thin-film axicons of larger diameter (1 cm range), needle beams of even much higher values (up to about 700:1) were obtained.

The spatial (temporally integrated) intensity distribution detected in a certain plane ($\Delta z = 1$ mm from the truncating aperture, corresponding to the measured Rayleigh range) for the same configuration is shown in Fig. 139. The symmetry and smoothness of the distribution function are a further proof for the nondiffracting propagation of the truncated Bessel-like beam, i.e. its property to be a high-quality "needle beam".

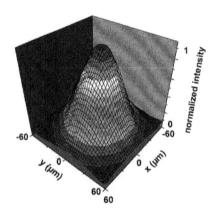

Fig. 139. Transverse intensity distribution of the truncated Bessel-like beam in Fig. 138 detected in a selected plane at the end of the Rayleigh range ($\Delta z = 1$ mm behind the truncating aperture).

In further experiments, *arrays* of up to 5 of such truncated Bessel-like beams were generated simultaneously by an array of holes of 30 µm diameter. The setup is schematically sketched in Fig. 140. Figure 141 shows the intensity distribution of the 5 sub-beams at a propagation distance from the truncating hole array of 1 mm. It was observed that the interference of neighboring nondiffracting beams starts to be visible far behind the typical distance of about 1 mm for a coupling by the diffraction of a plane wave at the edges of the 30-µm apertures (Fig. 142). Interference ripples of significantly high contrast appear only after a propagation path of > 4 mm. This multi-

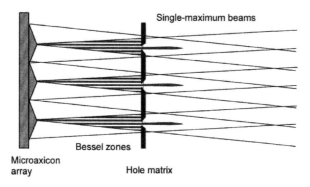

Fig. 140. Generation of multiple truncated Bessel-like beams by inserting an adapted hole array mask into the propagation path (axial position of the apertures: $\Delta z = 9$ mm, thin-film axicons: see data in Fig. 134).

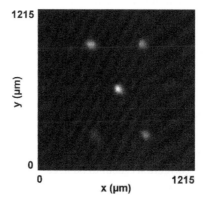

Fig. 141 Generation of multiple truncated Bessel-like beams by truncation with an array of 7 holes of adapted parameters (hole diameter 30 µm, period 405 µm). Because of the particular amplitude and wavefront profile of the incoming beam, only 5 of the beams were transmitted at detectable intensity, whereas 2 further spots on the right and left side appear to be very weak).

slit interference experiment clearly indicates the (pseudo-) nondiffracting nature of needle beams generated by self-apodized truncation, too.

It appears that a *Young's double slit experiment* with truncated Bessel beams should deliver very different results compared to the classical setup. If the interference experiment is repeated with focused ideal Gaussian beams (irrespective of the question how to generate an array of such foci), the outer parts of their intensity profiles will touch the edges and diffraction can not totally be excluded. Other specific advantages of truncated Bessel beams in comparison to focused Gaussian beams will be demonstrated in the frame of ultrabroadband spectral transfer, see paragraph 7.6.

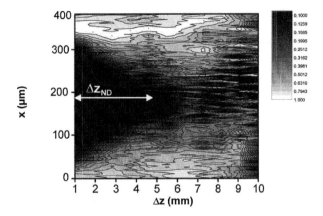

Fig. 142. Intensity interference pattern of an array of coherent microscopic nondiffracting beams after passing an array of truncating apertures in approximately self-apodizing operation mode (horizontal direction: propagation distance behind the apertures, vertical axis: transversal coordinate). The transversal spatial interval nearly corresponds to the distance between two beam centers (405 µm). Over a distance Δz_{ND} of > 4 mm ("nondiffracting" zone of propagation), no significant interference contrast is observed.

From the practical point of view it has to be mentioned that the accuracy of multichannel truncation depends on the number and size of channels because of local and global variations of the amplitude and phase of the incoming wave (inhomogeneities, deviations from plane wavefront). To overcome this drawback, additional *adaptive optical* corrections seem to be necessary.

Further progress might come from applying recently proposed *resonant structu*res (similar to those known from axicon laser resonators) which "recycle" the energy of the outer fringes otherwise wasted by truncation [P13].

6.3.6. Ultraflat thin-film axicons of extremely small conical angles as a second way to generate needle beams

With thin-film microaxicons of thicknesses in the range of 1-10 µm and diameters of 350-800 µm, Bessel-like beams with concentric fringes are generated [2000_6, 2001_7, 200_8]. With advanced deposition technologies and substrates of higher surface accuracy, even thin-film microoptics of structure depths well below 1 µm (*nanolayer microoptics*) were fabricated. Special features of nanooptical microoptics were already discussed in Chapter 4. On the basis of the dependence of the spatial frequency of Bessel fringes on the conical beam angle, axicons can be designed where the maximum beam diameter in the Bessel zone coincides with the radius of the first maximum of the

Bessel function. The realization of this particular case is the *second possibility to generate single maximum nondiffracting beams*. The principle is schematically drawn in Fig. 143 [2005_06, 2005_8, 2005_10, 2006_7]. In Fig. 144, the characteristic changes of the spatial fringe structure are illustrated by transverse intensity profiles in a plane at the center of the Bessel-like zones.

In experiments, arrays of microscopic needle beams were generated by transparent [2005_10] as well as reflective thin-film components at near infrared wavelengths [2005_10, 2006_6, 2006_7] and by transparent components at VUV wavelengths down to 157 nm [2003_9, 2004_1]. In Fig. 145, the transverse intensity pattern shaped from a Ti:sapphire laser beam at a pulse duration of 10 fs and a center wavelength of 790 nm is

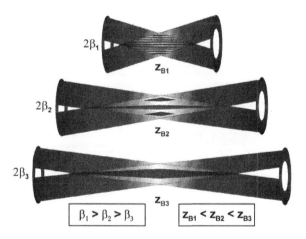

Fig. 143. Generation of single-maximum pseudo-nondiffracting beams ("needle beams") with axicons of ultra-small conical angles (schematically): stretching of the length of the Bessel-like zone and enhancement of the fringe period to smaller conical beam angles (from top to bottom).

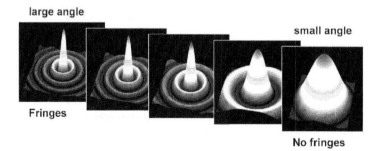

Fig. 144. Increase of the central spot diameter and reduction of the spatial frequency in the transverse profiles of Bessel-like beams in the center of the Bessel-like zone for stepwise reduction of the conical beam angles (schematically).

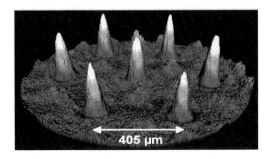

Fig. 145. Needle beam array generated from a 10-fs Ti:sapphire laser oscillator pulse by a hexagonal matrix of purely reflective nanolayer microaxicons of inverse-Gaussian thickness distribution (structure depth about 300 nm, period 405 µm, see design in Fig. 71c and structure in Fig. 74 in Chapter 4).

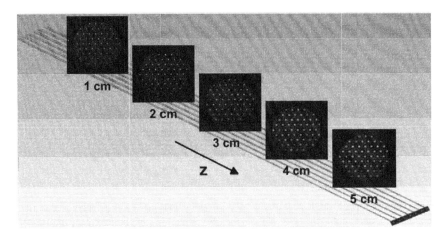

Fig. 146. Needle beam array generated from a 10-fs Ti:sapphire laser oscillator pulse by a hexagonal matrix of transparent Gaussian-shaped microaxicons (fused silica on silica, thickness 900 nm, period 405 µm). The Rayleigh length was determined to be 31 mm.

reproduced. The propagation of an array of extremely extended Bessel-like focal zones generated by sub-micron thick *transparent* Gaussian microaxicons (fused silica on silica, vertex height about 900 nm) from the beam of a 10-fs Ti:sapphire laser oscillator is demonstrated in Fig. 146. A Rayleigh length of 31 mm and an aspect ratio (here the ratio of the Rayleigh length to the minimum radius of the beam waist in the focal zone) was determined to be 694:1.

Both with Gaussian and inverse Gaussian nanolayer microaxicons, a high *robustness against angular tilt* was observed. This property will play an important role for the applications described in the next Section.

Finally, it should be mentioned that needle beams can also be generated by methods which reshape the angular distribution (e.g. nonlinear processes like the Kerr effect) or influence the shape and extension of effective interference volumina (e.g. spatial dispersion induced by travel time differences of different parts of ultrashort wavepackets [2004_21, p. 51]).

6.4. Shack-Hartmann wavefront sensing at extreme laser parameters with Bessel-like beams

6.4.1. Particular features of Shack-Hartmann sensors with Bessel-like nondiffracting beams

The principle of Shack-Hartmann sensors (SHS) was already previously introduced (see Chapter 2, in particular the schematic in Fig. 14b) [1995_16, 1995_17]. The detection of local wavefront tilts by long-focal-length microlens arrays is well known, e.g. from astronomical applications [1992_11]. Recently, the operation range of the classical Shack-Hartmann principle could be significantly extended by replacing the classical microlens arrays by *thin-film microaxicon arrays* [2005_9, 2005_06]. In this way, axicon-specific as well as thin-film-specific improvements were achieved. The advantages of *axicons* for Shack-Hartmann-type wavefront sensors briefly summarized [P12]:

- Pseudo-nondiffracting beams of *large depth of focus* are generated. Therefore, the system is less sensitive against axial displacement, and wavefronts of shorter radii of curvature can be characterized without suffering from leaving the focal zone at certain parts of the focal spot array [2002_6].
- In comparison to spherical lenses, axicons show a higher *robustness against angular tilt*. This was known for lenses with strong spherical aberrations and macroscopic axicon lenses [2000_14] and DOE axicons [2003_6]. Still, DOE are typically limited in efficiency and spectral transfer by their working principle, the diffraction.
- *Self-reconstruction* properties could benefit measurements in environments with significant distortions of the optical field (scattering, shading, absorption). The capabilities of advanced wavefront sensors in this way have still to be tested out.
- Bessel-like beam arrays carry not only wavefront information in the positions of the central spots but also in the *symmetry and spatial frequency distribution* of the fringe patterns.

The *combination* of these features with those of *thin-film structures* adds to the list of advantages some more useful items [2004_14, 2004_19, P12, P16]:

- Thin-film structures enable to work at extreme parameters of the light field (power, pulse duration, wavelength, bandwidth) because of *low-dispersion* design options (see Chapter 4). In particular, *reflective* setups can be realized. For the operation in reflection mode, both reflective layers and the tilt robustness of axicons are important.
- The capability to work at *short wavelengths* (VUV demonstrated [2004_1], X-ray possible after adapting mask-shadow techniques to well-established X-ray mirror fabrication technologies).
- Possibility to integrate *spectral* selectivity (e.g. by using graxicons or multilayer filters) and to realize *apodizing* elements by using multilayers (e.g. by GRMMA).
- Single-maximum beams (*needle beams*) can be realized (see the previous paragraph).

Further progress was recently obtained in the frame of ultrashort pulsed wavefronts where the Shack-Hartmann method was combined with the principle *autocorrelation* (see paragraph 7.3.).

Figure 147 shows the principle of wavefront sensing with tilt-tolerant axicon arrays (Shack-Hartmann sensor with arrays of Bessel-like beams) in comparison to the

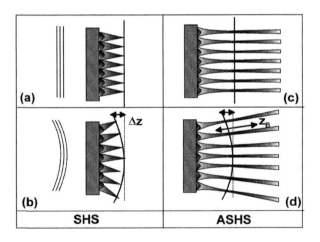

Fig. 147. Comparison of classical Shack-Hartmann wavefront sensor (SHS, a, b) and axicon-based Shack-Hartmann wavefront sensor (ASHS, c, d): (a) For a plane wavefront, all of the focal spots of SHS appear in the same plane. (b) For a strongly curved wavefront, however, the SHS focal spots leave the plane (Δz) so that they are enlarged and blurred by tilt-dependent aberrations. (c) The ASHS also works by the principle of SHS of splitting the beam into a beam array and focusing the sub-beams. (d) However, its is more tolerant against the curvature by exploiting the advantageous features of pseudo-nondiffracting beams (large axial extension of focal zones, robustness against tilt and defocus). z_B denotes the depth of the Bessel-like zone.

classical Shack-Hartmann setup. For a plane wavefront, the focal spots of SHS appear in one and the same plane. A strong curvature of the wavefront, however, leads to an axial shift of the focal spots to out-of-plane positions. Furthermore, the spots of spherical microlenses are enlarged even at relatively short distances from the focal plane and blurred by tilt-dependent aberrations. The axicon-based SHS (ASHS) is more tolerant against the wavefront curvature because of the large axial extension of focal zones and its robustness against tilt.

The working principle of SHS with Bessel-like beams is demonstrated by simulated transverse elongations in the image plane of a detector (Figs. 148a-c) [2004_21].

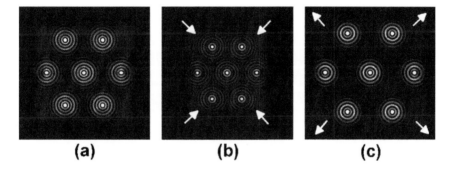

(a)   (b)   (c)

Fig. 148. Working principal of an axicon-based Shack-Hartmann wavefront sensor (ASHS) generating Bessel-like beams: Simulated patterns for (a) illumination with a plane wave, (b) a converging spherical wave with a radius of curvature of -0.05 m, (c) a diverging spherical wave of +0.05 m. The information on local wavefront tilt is encoded in the transversal shift of complete Bessel fringe systems instead of the shift of focal spots (schematically). The fringes contain additional information on astigmatism (symmetry).

6.4.2. Transmissive and reflective setups with angular-tolerant thin-film microaxicons

For applications in low-dispersion, reflective wavefront sensors, the most critical point is the *tilt robustness* of the micro-mirrors. Conical axicons show characteristic deformations of the Bessel-like fringe patterns as a function of the angle of incidence [2000_14]. The data of a Shack-Hartmann measurement are reliable over a certain range where the central spots are not erased by interference and the central region of the beam is not significantly affected by crosstalk (diffraction, reflection or refraction). For Gaussian-shaped microaxicons, a similar tilt dependence like in the case of conical lenses was found. It was further shown that the range of reliable operation can be extended by applying thin-film microstructures [2005_10]. Bessel fringe intensity patterns for microaxicon lenses of about 5.6 μm (average conical angle 2.25°) as a function of angular tilt and propagation path are depicted in Fig. 149. Obviously these particular structures can be used for wavefront sensing at incident angles of up to 25° and distances up to 15 mm.

Furthermore the reconstruction of wavefronts was tested with a *reflective setup*. An array of inverse-Gaussian-shaped micro-axicon mirrors (Au on ZnSe, hexagonal, period 405 µm, structure depth 300 nm) was illuminated at angles of incidence between 10° and 20°. Figure 150 shows the vector field of the transversal displacement of the centers of the needle beams in a certain plain. The corresponding reconstructed wave field can be found in Fig. 151.

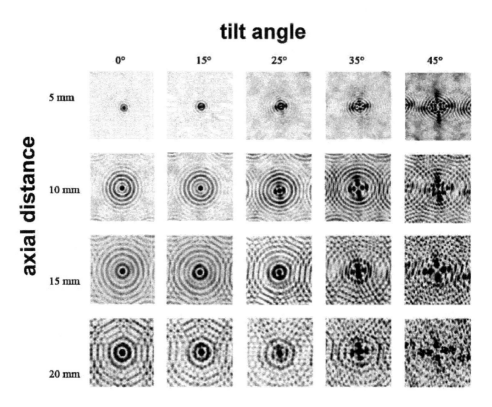

Fig. 149. Bessel-like fringe patterns generated by thin-film microaxicons of Gaussian thickness distribution (fused silica on silica, vertex height 5.6 µm) in dependence on tilt angle and axial distance (field of view: 460 x 460 µm²).

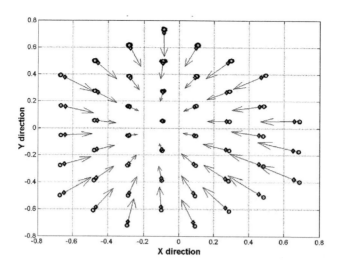

Fig. 150. Transverse shift of an array of focal spots of needle beams generated from a Ti:sapphire laser beam (pulse duration 10 fs, center wavelength 790 nm) in reflection (FOV 2.84 x 2.84 mm^2) after passing a transparent object (convex glass lens, f = 50 cm, distance from the array 11.8 cm). The shift vectors result from the comparison to a reference measurement without the object to compensate for the wavefront curvature of the measuring laser beam.

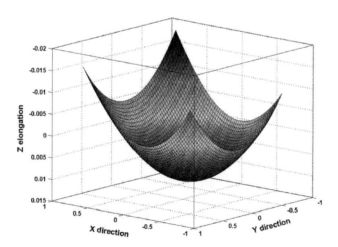

Fig. 151. Wavefront reconstructed on the basis of the data in Fig. 150 obtained by ASHS with needle beams in reflection setup.

6.5. VUV laser beam array generation and multichannel materials processing

6.5.1. VUV beam array generation with thin-film microoptics

Because of the typically low roughness (0.5-3 nm) and transmission down to the vacuum ultraviolet spectral range (minimum value of about 120 nm for MgF_2), thin-film microoptics and structures transferred into VUV-capable bulk material are favored

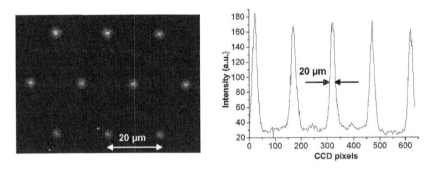

Fig. 152. VUV needle beams generated with ultraflat calcium fluoride microaxicons (light source: F_2 laser emitting at a wavelength of 157 nm, beam shaper: hexagonal array structures transferred from silica in calcium fluoride, final height 90 nm, period 405 µm). Left: part of the intensity pattern on a converter screen, right: cut at a distance of 8 mm from the axicons.

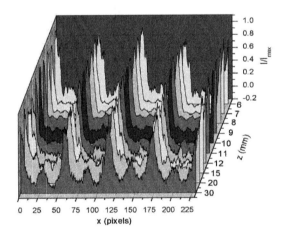

Fig. 153. VUV needle beams generated with composite layer structures (fused silica on calcium fluoride) microaxicons (light source: F_2 laser light source: F_2 laser, wavelength 157 nm, beam shaper: composite structures consisting of silica on calcium fluoride, height 2 µm, period 405 µm). At a distance of about 30 mm, additional interference ripples appear.

for short-wavelength applications [2003_9, 2005_9, 2006_6]. Figure 152 shows the intensity distribution of an array of *VUV needle beams* generated with Gaussian-shaped microaxicons transferred into CaF$_2$ (final height h_{max} = 90 nm), detected by imaging a converter screen onto a CCD camera. The propagation distance was 8 mm [2003_9].

With thin fused silica structures on CaF$_2$, Rayleigh lengths of up to 5 cm were obtained. Radial cuts through the propagating intensity pattern of such beam arrays can be found in Fig. 153. Here, the height of the fused silica microaxicons was 2 µm.

The experimental results were compared to theoretical simulations for single calcium fluoride elements based on the Rayleigh-Sommerfeld diffraction theory [2003_9]. In Fig. 154, the theoretical beam propagation is shown for a propagation path of 10 cm. By varying the height of the axicons from 100 to 600 nm (from left to right), the transition between needle beams and beams of Bessel-like fringe structure was

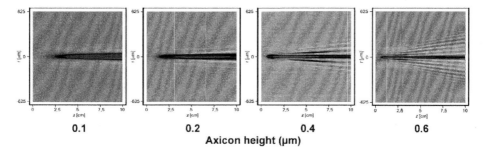

| 0.1 | 0.2 | 0.4 | 0.6 |

Axicon height (µm)

Fig. 154. Transition between needle beams and Bessel-like beams with fringes depending on the structure parameters (simulation for a wavelength of 157 nm, normal incidence, no absorption). The experimental results in Fig. 152 for a thickness correspond to the needle beam region (outer left picture).

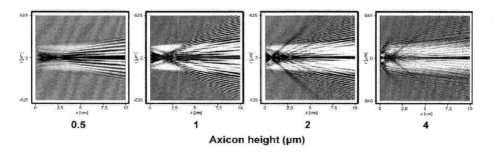

| 0.5 | 1 | 2 | 4 |

Axicon height (µm)

Fig. 155. Simulated change of the spatial structure of Bessel-like beams generated by composite axicon structures (fused silica on calcium fluoride, wavelength 157 nm, no absorption) depending on the height of the axicons. Towards higher thickness (from left to right), high-frequency distortions caused by diffraction at the edges appear.

found. The beams generated in the experiment with the 90-nm structures (Fig. 152) correspond to the *region of needle beams* (outer left side of Fig. 154). Composite structures of higher thickness (fused silica on CaF$_2$ substrate) are simulated in Fig. 155. One recognizes an increasing influence of *high-frequency spatial distortions*.

6.5.2. Beam cleaning by absorption

In the simulations concerning to Fig. 155, absorption was not taken into account. If the absorption is included in the model, however, important modifications are found [2004_06]. A fused silica layer with refractive index of $n = 1.684$ and extinction coefficient of $\kappa = 0.052$ at a wavelength of 157 nm assumed, a *beam cleaning* effect with respect to the spatial frequency content can be obtained. In Fig. 156, the spatially variable absorption of a typical axicon of a thickness of about 1.7 μm is plotted together with the thickness profile. The comparison of the simulated beam propagation with and without the contribution of layer absorption (Fig. 157) shows a modification of the interference fringes by the absorption. Higher spatial frequencies are filtered out so that a kind of beam cleaning is obtained. The central part (highest absorption) appears to be "self-apodized" by absorption (whereas the outer edges of the axicons are self-apodized by the vanishing thickness). The result agrees better with the observed beam structures then the calculations without absorption.

It can be concluded that the integration of well-designed absorbing layers in composite structures can be used to realize an amplitude mask function to modify or improve the beam structure of pseudo-nondiffracting beams. Beside the apodizing properties of ultraflat edges and the self-apodizing truncation setup, absorptive masking represents a further approach of self-apodization.

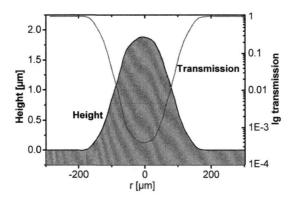

Fig. 156. Thin absorbing laser (fused silica) as attenuating mask (AM) structure at a wavelength of 157 nm. (Gaussian shaped microaxicon, extinction coefficient $\kappa = 0.052$). Left ordinate: height of simulated layer profile; right: corresponding transmission (logarithmic scale).

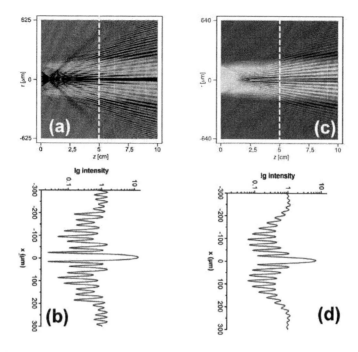

Fig. 157. Comparison of the simulated beam propagation of pseudo-nondiffracting beams in VUV without (a, b) and with (c, d) absorption (wavelength 157 nm, microaxicon: fused silica on CaF$_2$, absorption data see Fig. 156). The pictures on bottom show radial cuts corresponding to the positions at the dashed lines in the pictures on top. The absorption causes a filtering of the spatial frequency spectrum.

6.5.3. VUV materials processing with thin-film microoptics

The application of "diffraction-free" beams for lithographic applications was proposed in 1997 by Erdélyi et al. [1997_46]. Nearly at the same time, the first experiments in axial beam shaping [1997_8, 1997_9, 1997_10] and *multichannel materials processing* with non-spherical thin-film microlenses were successfully performed. With a high-power KrF-excimer laser beam, arrays of holes of unconventional shape were drilled in BK7-glass [1998_3]. The beam-shaping lenses were fabricated by depositing fused silica structures (of the type as shown in Fig. 60) with Gaussian and double-parabolic profiles on silica substrates. The simultaneous structuring of glass and polymer fibers with multiple extended focal zones of different types of lasers was demonstrated as well [1999_10].

Figure 158a presents schematically the laser treatment of bulk material with arrays of pseudo-nondiffracting beams of small conical beam angle (needle beams). With the help of a second lens, a scaling of the pattern written in the material can be obtained

(Fig. 158b). Arrays of holes were drilled by shaping and focusing a fluorine laser beam with calcium fluoride microaxicons. The interaction with materials like polymethyl methacrylate (PMMA), BK7 and silica was investigated. In PMMA, the structure depths of 0.2 - 2 μm and FWHM-diameters of 4 - 30 μm were obtained with 12 pulses of 8.5 mJ. A selected microscopic hole is shown in Fig. 159.

It was found that the major problem in materials processing with nondiffracting beam arrays is the uniformity of the illuminating beam. The demands on the spatial intensity stability are similar to the requirements on high-precision lithography. A

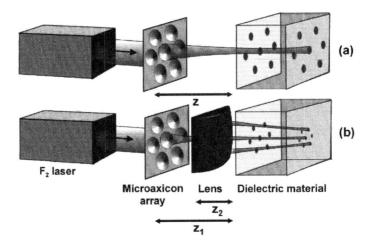

Fig. 158. Multichannel materials processing with arrays of needle beams without (a) and with (b) additional focusing optics for a scaling of the spot distance on the surface of the workpiece. The beam of a fluorine laser is shaped by an array of VUV-transparent thin-film microaxicons (schematically).

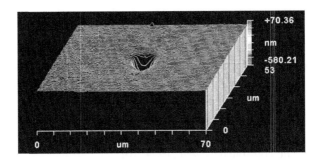

Fig. 159. Hole drilled in PMMA with a VUV-needle beam (wavelength 157 nm, 12 pulses, pulse energy 8.5 mJ) as a part of an array of beams. The dimensions of the hole are 5.6 x 6.6 μm² (interferometrically measured with $ZYGO^{TM}$ white light interferometer).

second problem arises from the homogeneity of the array structure. In particular at high intensities, small fluctuations of the intensity profile can exponentially be amplified by multiple-photon processes thus causing errors in the written structures.

Recently, another experiment was performed with linear arrays of *ultraflat cylindrical microlenses* of Gaussian thickness profile in one spatial direction. Here, arrays of *focal lines* (in transverse direction) instead of circular focal spots were generated. The setup is schematically drawn in Fig. 160. Figure 161 shows a part of the array of cylindrical microaxicon structures used as the beam array shaper. The material was MgF$_2$ directly deposited on MgF$_2$. The height of the axicons was 195 nm and the array period 500 μm. The shaped beams propagate as pseudo-nondiffracting beams in one spatial direction and are linearly extended in the other direction (arrays of "light

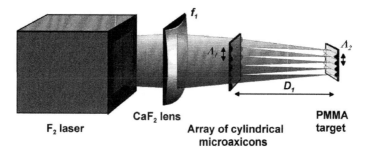

F$_2$ laser   CaF$_2$ lens   Array of cylindrical microaxicons   PMMA target

Fig. 160. Setup for the scalable multichannel materials treatment with cylindrical pseudo-nondiffracting beams (schematically). The beam of a fluorine laser is shaped by an array of cylindrical microaxicons (Gaussian profile in one spatial direction).

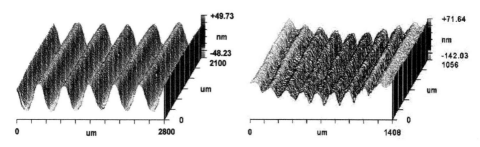

Fig. 161. Part of the array of cylindrical microaxicons used to shape an array of pseudo-nondiffracting beams with focal zones of linear shape in transversal direction in VUV at a wavelength of 157 nm (MgF$_2$ directly deposited on MgF$_2$ substrate, period 500 μm, structure height 195 nm).

Fig. 162. Apodized grating structure in PMMA simultaneously written with an array of cylindrical pseudo-nondiffracting F$_2$ laser beams (wavelength 157 nm, grating period about 150 μm). The thickness envelope function of the grating was obtained by exploiting the intensity envelope of the laser.

slices"). Long-period grating patterns of different periods were written in PMMA (optical multichannel structure transfer). By changing the distance between CaF_2 lens and target, the period of the written gratings could be scaled between 50 and 400 μm. A surface grating in PMMA is shown in Fig. 162. The thickness of the gratings could be modulated with an envelope function by exploiting the intensity envelope of the laser.

At this point, it may be instructive to compare the generation of Bessel beams with the standard configuration for the generation of *holographic gratings*. The situation is amazingly similar: a periodical intensity modulation is obtained by two interfering beams crossing at a certain angle. However, there is an important *difference*: The conical angle exists only in *one plane* so that the convergence is of two-dimensional instead of three-dimensional nature (the special case with two plane faces is well known as Fresnel's biprism). Therefore, instead of a central maximum surrounded by Bessel fringes one ends up with a grating of lines of equal intensity. Here, a similar setup was used so that each cylindrical microaxicon should generate its own light grating and the optical system should not be applicable. The trick was simply to reduce the thickness of the axicons so much that just a *single maximum* appears (analogous to the case of needle beams). This is the only one-element refractive way to the pseudo-nondiffracting light slice!

Chapter 7

SPATIO-TEMPORAL BEAM SHAPING AND
CHARACTERIZATION OF ULTRASHORT-PULSE LASERS

7.1. Motivation

In the previous Chapter it was demonstrated how the specific properties of thin-film microoptical beam shapers can be exploited to control the propagation of light in *spatial domain*. In modern optics, moreover, shaping and diagnostics of highly intense ultrashort wavepackets in *temporal domain* belong to the most challenging tasks. Pulses as short as a few or less cycles of the optical field and spectral bandwidths of up to one or more optical octaves have to be generated, amplified, measured, transformed, propagated and detected. To affect the laser performance, pulse propagation, and laser-matter-interaction, the beam control has to be extended to the *spatio-temporal domain*. In particular, the motivation for simultaneously tailoring and monitoring pulses in space and time arises from basic phenomena as well as their application potential:

- In few-femtosecond range, pulse travel time effects lead to *space-time coupling* (i.e. spatial and temporal parameters can not be separated [1999_26, 2002_48, 2002_49]). Space-time coupling plays an important role in stretching and compressing pulses in chirped pulse amplification (CPA), pulse shaping, coherent control, focusing and angular beam shaping.

- High intensities cause *nonlinear optical interactions* in the laser system itself, in optical components (mirrors, prisms, lenses, gratings) passed by the pulse, in air or on targets (e.g. Kerr effect). This causes spatial distortions as well.

- Temporal chirp and angular distribution are spatially modified by diffraction, absorption and dispersion (spatial chirp, angular dispersion) [2000_9, 2001_38, 2003_33, 2003_34, 2005_36]. These phenomena correspond to a space-variant *spectral phase* which is combined with *pulse front* deformations (in most simple case a *tilt*) and different *wavefront* curvatures.

- Ultrafast multichannel *materials processing* is of increasing interest [1999_25, 2001_37] because of the options of high-speed and addressable operation.

- By combining spatial and temporal localization, a more sophisticated *manipulation of matter* is enabled (e.g. trapping, guiding and accelerating of atoms or tweezing and rotating of particles and biological cells).

- Ultrafast *measuring techniques* for fast-changing mechanical or optical parameters (interferometry, holography, fringe projection) on the basis of structured beams and beam arrays promise applications beyond the limits of classical methods.
- Basic concepts of next generation devices for massive-parallel optical communication, data processing and storage will take advantage of *spatio-temporal and spatio-spectral multiplexing and encoding* (e.g. by programming a spatially dependent spectral phase). The free parameters in space, spectrum and polarization enable for a high information content in single ultrashort wavepackets. Low-dispersion time-to-space and space-to-time converters have to be designed [2004_5].

For all these reasons, the direct space-time characterization of ultrashort pulses as well as adaptive techniques for shaping pulses in space and time have to be regarded as key problems. Almost all of the known techniques for a spatio-temporal beam diagnostics are limited to one spatial coordinate and/or multi-shot regime (see, e.g., [2001_40, 2002_50, 2003_33, 2003_34]). In this Chapter, selected contributions of thin-film microoptics to the progress in this field will be addressed.

7.2. Coherence mapping

7.2.1. Microoptical approaches based on multichannel interferometry

To evaluate the spatial uniformity of ultrashort pulses and to read out spatially encoded temporal and/or spectral information, linear optical methods and integrating detectors can be applied [2002_48]. A mapping of the temporal coherence in space has to be performed with spatially resolving interferometric setups. The following approaches of a multichannel interferometry are all based on microoptical array components ("matrix processors" [2001_9, 2001_10, 2002_10]) (Fig. 163):

(a) Arrays of spatially variable Fabry-Pérot etalons generating interference fringes by intra-cavity interference (see Chapter 3) [2003_3]

(b) Arrays of microaxicons generating Bessel-like fringe patterns in free space (see Chapter 6) [P6]

(c) Arrays of phase and/or amplitude elements generating self-imaging (Talbot) patterns in free-space (see Chapters 1 and 6) [P5]

The clear advantage of *separated array zones* (a, b) is the elimination of errors by an overlap caused by wavefront curvatures. A disadvantage is the reduced spatial resolution (which could be improved by scanning and interpolation procedures). In *linear optical regime*, the analysis of the coherence distribution (without additional

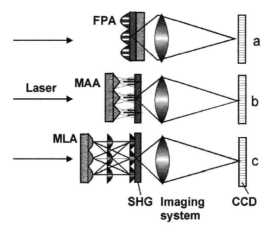

Fig. 163. Multichannel approaches for a spatially resolved detection of temporal coherence based on microoptical matrix processors (coherence mapping): (a) generation of intra-cavity fringes in a Fabry-Pérot-Array (FPA) of spatially variable elements, (b) free space interference generated by a micro-axicon array (MAA), (c) free-space Talbot effect generated by a microlens array. The setups in (b) and (c) can also be operated in reflection. SHG layers are introduced as nonlinear elements to derive temporal information.

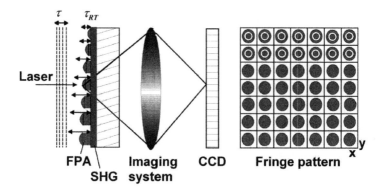

Fig. 164. Matrix processor for time-to-space conversion of a spatially uniform pulsed beam based on a non-uniform array of Fabry-Pérot etalons (schematically). The array consists of thin-film microlenses deposited on a nonlinear-optical layer for SHG. The temporal coherence τ is detected by the different etalon round trip times τ_{FP}.

measurements) delivers a map corresponding to the *spectral parameters* of the radiation. To derive *temporal information*, a nonlinear processing (e.g. by second or third harmonic generation) is necessary.

If, vice versa, the uniformity of the pulse can be taken for granted and a specific information is encoded in the temporal characteristics, the type-(a)-arrangement can be modified by integrating elements of different height (*non-uniform Fabry-Pérot-array*, schematically in Fig. 164) [P8]. In this case, the temporal coherence information is transformed into a global fringe contrast information (two-dimensional function over all elements in space) so that the processor works as a *time-to-space converter*.

The operation principle of coherence mapping with thin-film etalons was first demonstrated in a linear femtosecond laser experiment with a single Gaussian-shaped element (Fig. 165) consisting of fused silica on polycarbonate [2000_3, 2001_10]. The maximum layer thickness was > 9 μm.

Interference patterns were generated under three different spectral and temporal conditions: (a) cw laser diode emitting at a wavelength of 825 nm (as a reference), (b) Ti:sapphire laser at a center wavelength of 800 nm and a pulse duration of 26 fs, and (c) Ti:sapphire laser at a center wavelength of 800 nm and a pulse duration of 12 fs (Fig. 166). The pulses were generated in a Ti:sapphire laser system with chirped pulse amplification (CPA) in combination with an Ar-filled hollow waveguide [1997_47]. The relatively complex spectral distributions for the pulsed measurements are shown in

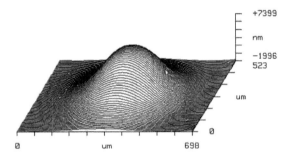

Fig. 165. Thickness profile of a single Gaussian-shaped thin-film etalon used for femtosecond interference experiments. The spectral bandwidth of an ultrashort pulse was tested by the interference contrast (SiO$_2$ layer on 0.5 mm thick polycarbonate substrate, refractive indices $n = 1.44$ and $n = 1.57$ at 800 nm, respectively; structure height > 9 μm).

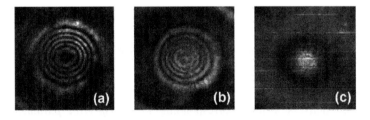

Fig. 166. Intensity fringes generated by intra-etalon spectral interference in the Gaussian-shaped layer structure corresponding to Fig. 167: (a) cw at 825 nm (laser diode), (b) 26 fs at 800 nm (Ti:sapphire laser, spectral FWHM > 40 nm), and (c) 12 fs at Ti: 800 nm (Ti:sapphire laser, spectral FWHM > 100 nm) [2001_10]. The repetition frequency of the laser was about 1 kHz.

Fig. 167. The intensity was attenuated by reflecting the beam at two prism surfaces in small incident angles. The vertex of the thin-film etalon was directed to the incoming beam and the fringe patterns were imaged with a microscope objective, a TV zoom lens and a CCD camera. The pulse durations were determined by autocorrelation. The results show the dependence of the interference fringes on the spectral bandwidth in agreement with theoretical simulations for polychromatic beams (compare to Chapter 3). Because of the softly changing profile, every part of the microstructure can be regarded as a wedge and the distance between neighboring fringes is approximately given by $\lambda_{CG}/2n$ (λ_{CG} - center of gravity of the spectrum, n - refractive index). With a gated camera, single shot measurements at 1 kHz repetition frequency of this laser system were possible.

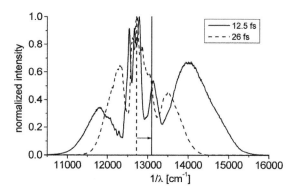

Fig. 167. Complex spectra for two different pulse durations of the CPA Ti:sapphire laser system (straight line: 12.5 fs, dashed line: 26 fs). The arrow indicated the blue shift of the center of gravity of the spectral distribution with shorter pulse durations.

7.2.2. Coherence mapping with thin-film Fabry-Pérot arrays

For a *two-dimensional spatial mapping* of interference contrast, the interference in high-resolution thin-film etalon arrays was used. The orthogonal matrix processor consists of resist microlenses (refractive index $n > 1.6$) on a silica substrate ($n = 1.46$) and has a period of 36 μm. By illuminating the array with 17-fs Ti:sapphire laser pulses at 1 kHz repetition rate, maps of first order interference patterns at the fundamental wavelength were detected as shown in Fig. 168a. The field of view was about 500 x 500 μm². The contrast of the fringes is visualized by the *cooccurrence matrix* (Fig. 168b) [1995_35, p. 11]. The cooccurrence matrix is a measure for the frequency of the occurrence of a combination of gray values of pairs of pixels at given distances.

The measurements were repeated with *nonlinear frequency conversion*. To generate the second harmonic (SHG), a thin BBO crystal was placed in close proximity to the

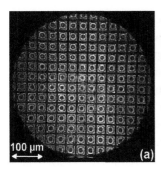 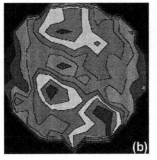 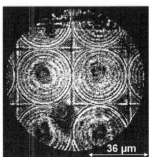

Fig. 168. Two-dimensional coherence mapping with orthogonal arrays of Fabry-Pérot etalons (spatial period 36 μm, > 130 elements completely involved in the field of view): (a) intra-etalon interference fringes, (b) cooccurrence map based on the fringe contrast data. The result is a visualization of the spatial inhomogeneity of the (temporally integrated) laser pulse spectrum.

Fig. 169. Two-dimensional coherence mapping with orthogonal arrays of Fabry-Pérot etalons (spatial period 36 μm, circular elements).

etalon array. However, the nonlinear crystal had to be tilted against normal beam incidence because of the phase matching conditions. Therefore, the distance between surface and NLO crystal in one spatial direction was a function of the transversal position and it was difficult to obtain sharp fringes over the complete field of view. In Fig. 169, second order fringes (detected with SHG in a 10 μm thick BBO crystal) of a small section of an array of etalon lenses is shown.

The performance of those experiments was further improved with a-axis ZnO nanolayers as nonlinear converters. As it was recently demonstrated, such layers enable for an efficient SHG even at *normal incidence* [2005_13, 2005_33, 2006_25]. Another approach is to generate both interference and SHG in a structured ZnO nanolayer and to integrate such *nonlinear functional nanolayers* into *planar-optical processor concepts*.

7.2.3. Coherence mapping with arrays of Bessel-like beams

A two-dimensional coherence mapping was also obtained with arrays of *Bessel-like beams* generated by thin-film microaxicons. The temporal and/or spectral information here is encoded in the envelope of the second order or first order fringe contrast (with or without SHG) and corresponds to second or first order *non-collinear autocorrelator*, respectively (see paragraph 7.3.3.). The linear response of Bessel-like beams to changes of a pulse spectrum can be well recognized in a logarithmic presentation as shown in Fig. 170. Here, the theoretical distribution for the square of the zeroth order Bessel function J_0^2 is compared to measured intensity fringes at two different pulse durations

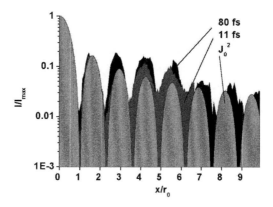

Fig. 170. Interference pattern of Bessel-like fringes in the plane of best contrast for different pulse durations of a Ti:sapphire laser (normalized logarithmic intensity, r_0 - radius of the central lobe of the Bessel distribution, x – radial coordinate).

(80 fs, 11 fs). The beam was generated with a Gaussian-shaped microaxicon (fused-silica on silica, thickness 5.7 μm, waist radius 150 μm).

With *arrays* of thin-film microaxicons, the temporal coherence can be mapped in two dimensions [P6]. The schematic setup of such a *Micro-Axicon-MAtrix processor* (MAMA) with a low-dispersion reflective microaxicon array is depicted in Fig. 171 [2001_7, 2005_11]. Experimental results of 2D fringe mapping are shown in Figs. 172 - 174. In the first figure, the interference patterns of a cw laser diode beam at a wavelength of 825 nm (Fig. 172a) and a 8-fs pulse of a Ti:sapphire laser (b) are

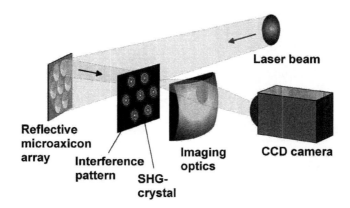

Fig. 171. Schematic of two-dimensional spatial mapping of temporal coherence with nonlinear multichannel non-collinear autocorrelation (cf. 7.3.3.). Reflective setup with low-dispersion microaxicon mirrors. For second harmonic generation, a thin BBO crystal or nanocrystalline ZnO layer is applied. The interference pattern is imaged onto a high-resolution CCD-camera.

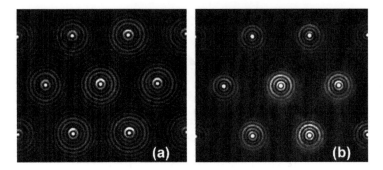

Fig. 172. Two-dimensional first order coherence mapping with arrays of Bessel-like beams: interference patterns generated by (a) cw beam at a wavelength of 825 nm, (b) 8-fs pulse of a Ti:sapphire laser (overexposed to enhance the visibility of weak outer fringes).

compared. To enhance the visibility of the weak outer fringes, the images were overexposed. A reduced number of fringes as well as reduced modulation depth at short pulses were found in good agreement with theoretical predictions, see [2004_21, p. 51]). Figure 173 shows the multichannel interference pattern over a larger array of Bessel-like beams (> 50 channels) obtained at a pulse duration of 10 fs. The SHG (center wavelength 395 nm) was generated in a high-efficiency, low-dispersion, ultraflat ZnO-nanolayer [2005_13, 1005_33, 2006_26] as frequency converter.

The contrast distribution of an 8-fs wavepacket (corresponding to 3 temporal oscillations of the optical field) detected over an area of 1.62 x 1.62 mm^2 in second-order operation mode (with SHG) is shown in Fig. 174. The fringe contrast with respect

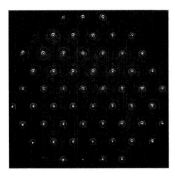

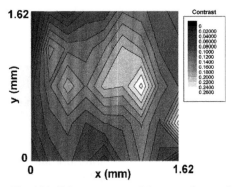

Fig. 173. Second order 2D-coherence mapping of 10-fs pulses with a large hexagonal array of Bessel-like beams generated by thin-film micro-axicons (period 405 µm). In the plane of best fringe contrast (axial distance $z = 9$ mm), a 300-nm thick ZnO film for SHG was placed.

Fig. 174. Coherence map of the central part of a Ti:sapphire laser pulse measured with an array of Bessel-like beams at a pulse duration of 8 fs.

to the m-th maximum of a Bessel-like distribution was defined by

$$C(x, y, m) = \frac{I_{\max}(m) - I_{\min}(m)}{I_{\max}(m) + I_{\min}(m)} \tag{120}$$

where x and y denote the coordinates of the channel centers, $I_{\max}(m)$ the average intensity of the m-th maxima and $I_{\min}(m)$ the average intensity of the directly neighboring minima. In the figure, the contrast map of the central maximum ($m = 1$) of a 8-fs Ti:sapphire laser oscillator pulse is plotted. The spatial contrast appears to be asymmetric because of distortions in laser and optical guiding system.

Particular *advantages* of using Bessel-like nondiffracting beams for coherence mapping with matrix processors are that

- the axial position of the nonlinear frequency converter is less sensitive against axial position and tilt,
- a low-dispersion (i.e. reflective) system design is enabled,
- and the beams are self-reconstructing.

Furthermore, the measurements with a MAMA processor deliver also angular information by working as an axicon-based Shack-Hartmann sensor (ASHS, see Chapter 6). The experimental setup of a matrix processor is shown in Fig. 175.

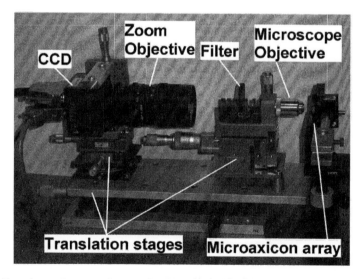

Fig. 175. Experimental setup of a matrix processor with CCD camera (here: Basler, 1 Mio pixels), zoom lens, microscope objective and microaxicon array. CCD-camera and objectives are translated simultaneously. For frequency conversion, SHG crystals or layers are placed between microscope objective and thin-film axicon array (not shown).

7.2.4. Decoding of axial coherence information with arrays of Bessel-like beams

The matrix processor technique described above (paragraph 7.2.3.) can also be used to detect spectral and/or temporal information encoded in the *axial propagation characteristics* of an ultrashort-pulsed beam. The example of a microscopic Bessel-like beam (conditions as in Fig. 174) is shown in Fig. 176 [2002_10]. The central maximum and 5 fringes (m = 1 - 6) were found to propagate with a distance dependent contrast function. The fringe with m = 6, e.g., has a distinct maximum at a distance of 9 mm from the generating thin-film microaxicons.

In general, an axial inhomogeneity can result from interaction of a highly intense laser pulse with matter or be generated in a transversal-to-axial transforming optical system (e.g. statically by graded multilayers or gratings, addressable by a spectrally selective SLM, or radially structured illumination of axicons). The principle might also be of interest for advanced tasks of metrology.

The data in Fig. 176 were obtained by sequentially imaging different planes of interest (axial scan). The simultaneous detection of the contrast at different depths (*3D contrast mapping*), however, is still an unsolved technical problem.

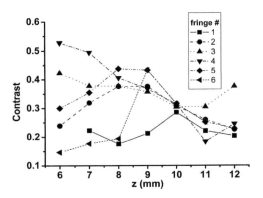

Fig. 176. Axial contrast information of the central maximum and 5 fringes (m = 1 - 6) of a propagating miniaturized 8-fs Bessel-like beam over the depth of the Bessel zone of about 6 mm. In this example, different fringes show maximal contrasts at different axial positions.

7.2.5. Coherence mapping with the Talbot effect

A third way (see Fig. 163c) to decode spectral and/or temporal information from the spatial characteristics of ultrashort pulses is to analyze self-imaged patterns [2001_9, 2002_10, P5] appearing at discrete distances called *Talbot planes* (see Chapters 2 and 6). The self-imaged array of focal spots generated by a microlens array of a pitch of 36 µm in the first Talbot plane with respect to the center wavelength (here: 790 nm) can be found in Fig. 177. The contrast patterns within a Talbot plane change with the spectral and/or temporal characteristics of the pulse so that an evaluation of the spatial homogeneity of the pulse is possible. Figure 178 shows the basic geometrical relations for the 1st Talbot plane. It is illustrated how the polychromatic self-image is split in axial direction and how the spatial resolution is reduced by a cross-talk between neighboring unit cells. Figures 179a -d show coherence maps for the 1st and 4th Talbot

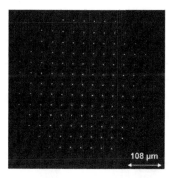

Fig. 177. Self-image of an array of focal spots of a 10-fs pulse generated by an array of microlenses of 36 µm pitch in the 1st Talbot plane (center wavelength 790 nm, transversal cut trough the intensity distribution).

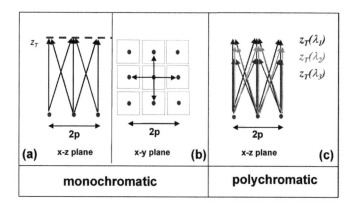

Fig. 178. Geometrical relations for the 1st Talbot plane: (a) x-z-plane and (b) x-y-plane for monochromatic, (c) x-z-plane for polychromatic illumination (schematically). In monochromatic case, only a single Talbot plane appears (a), whereas for polychromatic fields each spectral contribution (weighted by the spectral envelope function) is self-imaged to another plane (c). In the 1st Talbot plane, the spatial information is (in a simplified picture) dissipated over an area of 9 times the unit cell (i.e. the area of one square pitch in an orthogonal array). The corresponding transversal resolution is reduced by a factor of 3.

planes derived from self-imaging measurements with (a, c) cw beam and (b, d) a pulsed beam (Ti:sapphire laser), respectively. Because of the spectral dependence of the Talbot effect, the time-to-space conversion of ultrashort wavepackets with the Talbot effect leads also to an encryption of spectral and/or temporal information into the *axial* characteristics of the beam. The average intensity contrasts from these measurements are listed in Tab. 3. As expected, the values decrease with the pulse duration as well as the number of the Talbot plane. The contrast is significantly amplified in the 4th Talbot plane.

	average intensity contrast	
	1st Talbot plane	4th Talbot plane
cw	0.94	0.94
9 fs	0.82	0.57

Tab. 3. Comparison between average intensity contrasts in two different Talbot planes for a cw laser and a 9-fs Ti:sapphire laser pulse.

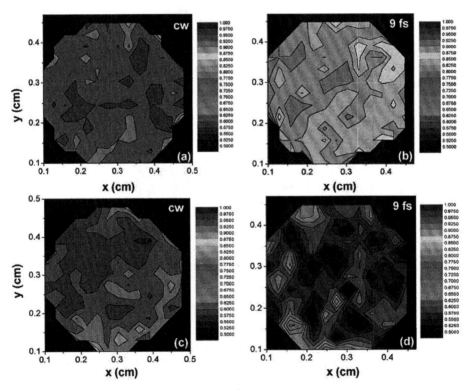

Fig. 179. Coherence map of the 1st (a, b) and 4th (c, d) Talbot plane determined by self-imaging measurements with a cw laser (a, c) and with a 9-fs Ti:sapphire laser beam (b, d). The contrast differences in the 4th Talbot plane are amplified compared to the 1st Talbot plane.

It should be mentioned that similar mapping and multiplexing techniques can also be developed on the basis of the closely related *Montgomery effect* [1967_1, 1968_1, 2003_23, 2003_24, 2005_26].

7.3. Spatio-temporal autocorrelation

7.3.1. Processing and characterization of ultrashort optical pulses

During the last decade, increasing interest in *processing and diagnosing* ultrashort pulses arose from the access to commercial femtosecond laser sources as well as from new applications in information technologies and laser-matter interaction [2004_5].

Various linear and nonlinear approaches of ultrafast optical processing were studied so that only selected examples can be addressed here. Three-wave-mixing or four-wave-mixing were applied to real-time encoding of phase information and time reversal of waveforms [1999_5, 2000_37]. Spectral information inversion by mixing four spectrally decomposed waves exploits the ultrashort response time of cascaded second-order nonlinearities [2000_38]. Packet detection in time-domain communication by interferometric first-order cross correlation [1993_26, 1999_27] and differentiation of ultrashort pulses [2006_32] were demonstrated. Information encoding in the temporal-frequency content of ultrashort pulses can be obtained by three wave mixing [1997_25]. Here, both amplitude and phase of a femtosecond laser pulse are derived simultaneously.

A first information about the temporal characteristics of very short signals can be obtained with the well-known methods of *intensity autocorrelation* [1985_7] and *interferometric autocorrelation* [1996_5, 1998_6]. Because of specific limitations of autocorrelation (spectral phase data can not completely be derived), so-called *"complete"* characterization methods were developed. The most important of these techniques are frequency-resolved optical gating (FROG) [1993_27, 2000_9, 2003_33], spectral phase interferometry for direct electric-field reconstruction (SPIDER) [1998_18, 2001_40], measurement of electric field by interferometric spectral trace observation (MEFISTO) [2005_38] or GRENOUILLE (grating-eliminated no-nonsense observation of ultrafast incident laser light E-fields) [2001_38, 2003_34]. A further, spectrometer-less method is the angular-resolved autocorrelation which yields a *chronocyclic* representation of the pulse by analyzing the divergence and curvature of SHG [1995_8]. Some of these techniques are well-established so far and deliver impressing spectral phase data. With ever decreasing pulse durations and increasing intensities, however, the spatial information of ultrashort pulses is of increasing importance as well [2004_10, 2005_36]. The light propagation through optical guiding systems and temporal beam shapers with space-to-time converting SLM also causes spatio-temporal distortions [1996_34].

As already mentioned, the majority of the known techniques is working without any spatial resolution or with a spatial resolution in only a single dimension (e.g. SPIDER [2001_40], spectral and spatial shearing interferometer [2002_50], particular types of FROG [2003_33, 2003_34]). Spatial resolution is possible with light-in-flight holography [1990_12] but here one has the problem to generate a high-quality reference wave of a polychromatic beam or to work sequentially with spectral filters [2004_10].

A *fully 2D spatially resolved pulse characterization* can be obtained by combining relatively simple autocorrelation techniques with low-dispersion thin-film microoptical beam shapers in novel system architectures as we will show in the following.

7.3.2. Concept of the collinear matrix autocorrelator based on arrays of Bessel-like beams

In previous Sections, single-shot multichannel processors for a coherence mapping were presented. In Chapter 6 it was shown that the specific features of thin-film components, axicon beam shapers and non-diffractive beams can be combined to improve the performance of wavefront sensors (axicon-based Shack-Hartmann sensor, ASHS). An alternative approach for a *"more complete"* pulse characterization with respect to a fully two-dimensional *spatial resolution* is to extend the ASHS principle by combining it with the collinear temporal autocorrelation.

For an interferometric autocorrelation of ultrashort laser pulses, the wavepacket is split by a beam splitter into two replica which circulate in an interferometric setup (e.g. Michelson or Mach-Zehnder). The delay between both pulses is tuned by shifting one of the arms of the interferometer. The first-order interference signal I_1 as a function of the delay time τ is measured by a time-integrating detector so that it is proportional to the squared electric field averaged over an oscillation period [1996_5, pp. 45-50; 1998_6, pp. 177-201]:

$$I_1(\tau) = \int_{-\infty}^{\infty} \left| E(t) + E(t-\tau) \right|^2 dt \ . \tag{121}$$

In the ideal case (balanced interferometer with infinitely thin beam splitter, nondispersive components), the signal is proportional to the first-order *autocorrelation function* $G_1(\tau)$ of the electric field:

$$I_1(\tau) \propto 2 \int I(t)dt + 2G_1(\tau) \tag{122}$$

or

$$G_1(\tau) \propto \int_{-\infty}^{\infty} \left| E(t) + E(t-\tau) \right|^2 dt \ . \tag{123}$$

$G_1(\tau)$ is a *symmetric function*. Therefore, any asymmetry information is lost in such a configuration. To approximate a complete description of $E(t)$, higher orders of autocorrelation $G_n(\tau)$ have to be determined via multiphoton processes of appropriate orders. A second-order autocorrelation measurement can be realized with second harmonic generation if the beam distortion introduced by the nonlinear converter can be neglected (thin crystal or layer). The corresponding intensity is

$$I_2(\tau) = \int_{-\infty}^{\infty} \left| \left[E(t) + E(t-\tau) \right]^2 \right|^2 dt \ . \tag{124}$$

For second order autocorrelation $G_2(t)$ follows:

$$G_2(\tau) \propto \int_{-\infty}^{\infty} \left| \left[E(t) + E(t-\tau) \right]^2 \right|^2 dt \ . \tag{125}$$

In a matrix processor with interferometric autocorrelation, the beam is analyzed in a large number of separated channels without cross-talk so that an array of autocorrelation data is obtained. Each autocorrelation function represents a unit cell of the matrix. The concept of such a matrix autocorrelator [2001_7] is shown in Fig. 180. The setup is similar to the single-shot matrix autocorrelator used for coherence mapping (compare to Fig. 171).

Here, the incoming ultrashort wavepacket is split by a beam splitter into two copies which are delayed with respect to each other as in the classical interferometric autocorrelator arrangement. In the schematic picture, a Michelson-type interferometer is drawn one mirror of which is translated by a piezo actuator. The two copies of the

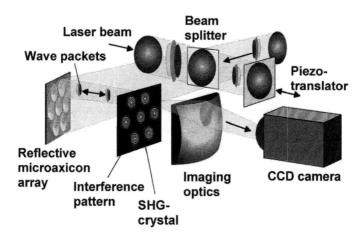

Fig. 180. Matrix processor for spatio-temporal diagnostics of ultrashort wavepackets based on a multichannel autocorrelator with Bessel-like beams (schematically).

wavepacket pass a thin-film microaxicon array (here: a reflective one) so that they are transformed into an array of pseudo-nondiffracting beams. The plane of interest is now imaged by a microscope objective onto a time-integrating CCD-camera. For nonlinear frequency conversion, a thin SHG crystal or layer is placed in the imaged plane. The important difference to the case of non-collinear coherence mapping is that not only transversal fringes but the axial interference pattern of these propagating fringe arrays is analyzed and a three-dimensional autocorrelation data set (x,y,t) is obtained. Instead of the temporal second order autocorrelation in eq. (125), a *spatio-temporal autocorrelation function* $G_2(x,y,\tau)$ is determined [2003_1]:

$$G_2(x,y,\tau) \propto \int_{-\infty}^{\infty} \left| [E(x,y,t) + E(x,y,t-\tau)]^2 \right|^2 dt \ . \tag{126}$$

It has to be noted that the conical beam angles in such a setup lead to specific small errors in the outer parts (in radial direction) of the fringes because of a weak spatial walk-off [2004_19]. Here, one takes advantage of the very small angles which can only be obtained with thin-film microaxicons.

In contrast to autocorrelation measurements of the complete beam with matrix cameras, the separation in pseudo-nondiffracting sub-beams eliminates the errors caused by a *curvature* of the wavefront (no overlap). With THG crystals as nonlinear frequency converters, the third order autocorrelation function can be derived. In this case, even asymmetry properties of the pulse are revealed. With triple autocorrelators, pulse asymmetry can also be detected. The more complicated interferometric setup comprises two delay lines and three pulse replica [2000_9, pp. 81-82].

7.3.3. Transversal autocorrelation information in Bessel-like beams

In addition to the axial interference, the autocorrelation information encoded in the *transversal direction* can be analyzed. It has to be taken into account, however, that the interference signal of a Bessel-like beam structure (concentric fringes) is basically *different* from the intensity pattern of a conventional autocorrelator. The relation between classical radial autocorrelation function $G(r)$ and the autocorrelation function of a Bessel-like beam generated with a *conical* axicon $G_{Bessel}(r)$ is [2004_21, p. 72]

$$G_{Bessel}(r) \propto \frac{G(r)}{r^2} \tag{127}$$

(r - radial coordinate). Using the relationship between r and the travel time difference τ

$$\tau = \frac{2r\sin\alpha}{c} \tag{128}$$

(a - axicon angle, c - light velocity) it can be shown that

$$G(\tau) \propto \tau^2 G_{Bessel}(\tau) \ . \tag{129}$$

If Gaussian-shaped axicons or other initial phase distributions are applied, the data analysis is more complicated (see [2004_21, pp. 74-76]).

7.3.4. Wavefront autocorrelation experiments

Apart from the spatially resolved temporal information, a matrix autocorrelator with Bessel-like beams delivers information on the *angular spectrum* of the pulse just like a Shack-Hartmann wavefront sensor [2002_6]. The spectrally averaged and temporally integrated wavefront information can be extracted together with the autocorrelation information so that the combined measuring technique was referred to as *"wavefront autocorrelation"* [2003_1, 2004_8, 2004_13, 2004_19, 2005_11]. The system geometry of a reflective device used in first wavefront autocorrelation experiments is illustrated in Fig. 181. Laser sources were a Ti:sapphire laser oscillator (pulse duration 10 - 12 fs, center wavelength 790 nm, repetition rate 75 MHz) and an amplifier (30 fs, 1 kHz).

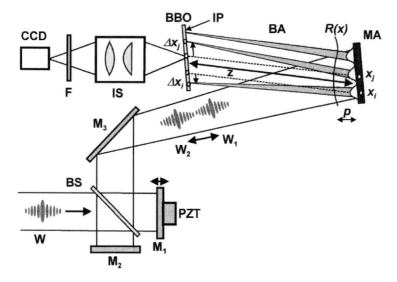

Fig. 181. System geometry for wavefront autocorrelation in a reflective setup (W - wavepacket, M_1 and M_2 - HR mirrors of the Michelson interferometer, PZT - piezo translator, BS - beam splitter, M_3 - HR mirror, MA - microaxicon array, W_1 and W_2 - replica of the wavepacket, BA - Bessel-like beam array, IP - interference pattern, BBO - Beta-barium borate nonlinear crystal, IS - imaging system, CCD - matrix camera).

Two collinear beams were generated in a Michelson interferometer with a 1-mm thick beam splitter coated with a broadband dielectric 50%-mirror. The oscillator beam had a diameter of 3 mm in the entrance plane after propagating 1.8 m in air. The diameter of the amplifier beam was 7 mm after an OPD of 2.3 m. Different hexagonal arrangements of Gaussian-shaped microaxicons were tested as beam shapers: (a) transparent fused silica structures on silica at normal incidence, and (b) pseudo-reflective axicons made of fused silica on Ag.

For second order autocorrelation, a 10-μm thick BBO-crystal was inserted. First and second order interference patterns were selected by color filters and imaged onto a CCD camera (Basler A101P, 1300x1030 pixels of 6.7 x 6.7 μm^2) by a combination of an achromatic microscope objective (4x) and a zoom lens. Nearly symmetric second order Bessel-like sub-beams (Fig. 182a) were generated using pseudo-reflective axicons (height 0.5-2 μm, period $p = 405$ μm) at an incident angle of $\alpha = 14°$ and an axial distance of $z = 10$ mm. At larger incident angles, the Bessel-like beams degenerate and Mathieu-like beams [2000_20] appear. This effect can be seen in Fig. 182b for an angle of 25° at a distance of $z = 14$ mm.

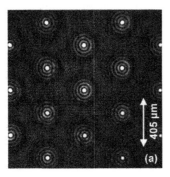 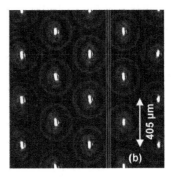

Fig. 182. Intensity patterns of second order Bessel-like (a) and Mathieu-like beam arrays (b) generated by hexagonal transparent and pseudo-reflective microaxicon arrays (period 405 μm). The incident angles were 14° and 25° and the distances between axicons and the center of the imaged plane were 10 mm and 14 mm, respectively.

The array pitch p pitch serves as a scale for the self-referential determination of local wavefront tilt. The local radius of curvature $R(x)$ of the wavefront for a transverse dimension x is determined from the displacements Δx_i and Δx_j of adjacent beams relative to their plane-wave positions x_i and x_j in the plane of axicons (see Fig. 180):

$$R(x) = \frac{z \cdot p}{(\Delta x_i - \Delta x_j)} \tag{130}$$

with the averaged transverse coordinate

$$x = \frac{x_i + x_j}{2} \; . \tag{131}$$

The transversal resolution δx given by pixel size and zoom factor was estimated to be 3.1 μm for 4x magnification. The resolution for the determination of the radius of curvature results from the working distance z and the period p and was found to be $\delta R \propto zp/\delta x$. Improvements of the sensitivity are possible by enhancing magnification and pixel resolution, by sub-pixel interpolation and additional scanning.

To visualize the data of 3D spatio-temporal correlograms, *2D representations* can be used. These are obtained by projections on the x-τ plane resulting from concatenating linear cuts through the 3D data. In Fig. 183, second order autocorrelation traces of a wavepacket without (Fig. 183a) and with (Fig. 183b) induced wavefront distortion are plotted. The distortion was caused by an $f = 50$ mm plano-convex glass lens at a distance of 118 mm from the array center ($\alpha = 25°$).

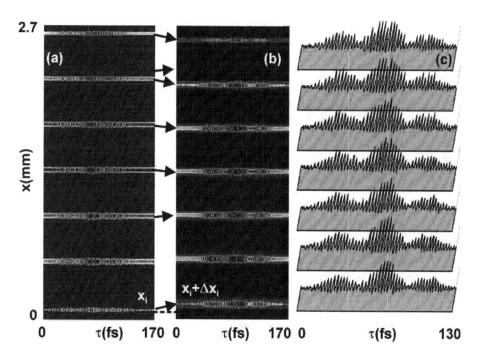

Fig. 183. Second order spatio-temporal wavefront autocorrelation with Ti:sapphire laser amplifier beam (pulse duration about 30 fs). Here, the wavefront is derived from comparing the correlogram traces detected with a phase object (lens) (b) to a reference data set (a). In (c), axial cuts through the centers of the corresponding intensity distributions are plotted.

By analyzing the *vertical* positions of nearest neighbors (unaffected by oblique incidence), the displacement of in total 52 maxima was determined. The calculated average radius of curvature $R(x) = 37.5$ cm agrees well with the assumed value of 38.2 cm for a perfect $f = 50$ mm lens. Cuts along the τ-axis in Fig. 183c correspond to the interferometric autocorrelation patterns which are known from spatially integrated measurements. Each of them represents a separate spatial channel of the matrix autocorrelator. The field cycles of the 30-fs pulse were clearly resolved (period: 2.4 fs at 790 nm, temporal resolution by the step width of the piezo: 0.34 fs). In Fig. 184, a 3D presentation of a part of an array of 3-cycle Bessel-like wavepackets is given (see also the colored cover picture).

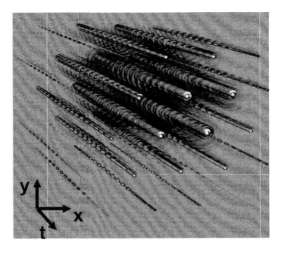

Fig. 184. Spatio-temporal autocorrelation of a few-cycle pulse of a Ti:sapphire laser detected by a matrix autocorrelator. The picture shows a part of an array of Bessel-like pseudo-nondiffracting beams at a pulse duration of about 8 fs and a center wavelength of 790 nm.

After carefully analyzing the spatio-temporal autocorrelation data according to the centers of the correlograms for each trace, subtracting intensity background and correcting for errors of the detecting system, the pulse duration can be represented as a *two-dimensional map* in space coordinates [2004_19]. The temporal characteristics of a Ti:sapphire laser oscillator pulse as a function of spatial coordinates is shown in the 2D map in Fig. 185. The central part of the beam was diagnosed over an area of 2.18 x 2.38 mm^2 corresponding to 31 separated sub-beams.

The beam center was found to be relatively uniform concerning to the temporal parameters. An average FWHM of the pulse length of 10.4 fs was calculated in excellent agreement with independent measurements. The standard deviation was well below 3%. This result means that

- the beam has a sufficiently good quality to be used in further amplification or processing stages,
- the detection technique has a sufficiently high sensitivity to resolve spatially encoded temporal and/or spectral information (spatio-temporal chirp) with intensity variations in few-percent-range.

With reflective thin-film microaxicons [2004_15] and low-dispersion nonlinear nanolayers (ZnO) [2003_10, 2004_7, 2005_13, 2005_33], an extension of the wavefront autocorrelation technique to even shorter pulses is enabled. Restrictions arise from the spectral response and transfer functions (including detectors, nonlinear converters and air turbulence). In the range of octave-spanning or broader spectral bandwidths, the spectral overlap of fundamental and second harmonic requires a reliable discrimination of the fundamental. Therefore, techniques based on a spatio-spectral or polarization selectivity or nonlinear gating have still to be developed.

In consideration of the encouraging first results of two-dimensional wavefront autocorrelation, the combination of the Shack-Hartmann sensor principle with one of the established *"complete"* characterization methods for the spectral phase like FROG or SPIDER seems to be a very promising task for the future.

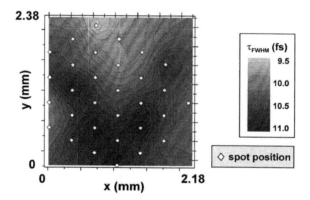

Fig. 185. Spatial map of the pulse duration of a 4-cycle femtosecond pulse from a Ti:sapphire laser oscillator. The array of 31 pseudo-nondiffracting needle-beams was generated by reflective thin-film microaxicons (gold on ZnSe, for the structures and beam profiles see Figs. 74 and 145). Second order autocorrelation was obtained in a 10-μm thin BBO crystal. An average pulse duration of 10.4 ± 0.3 fs was determined [2004_19].

7.4. Hyperspectral sensing of polychromatic wavefronts

7.4.1. Prospects for a spatially-resolved spectral phase measurement

The reconstruction of *polychromatic wavefronts* from ultrafast laser pulses is of increasing interest for several reasons:

- Reliability and scope of application of classical, spectrally averaged time-integrated wavefront sensing have to be proved.
- The extension of Shack-Hartmann-based autocorrelator techniques toward a spectral resolution would enable for a direct detection of *spectral phase fronts* of ultrashort-pulsed polychromatic wave fields.
- In this way, important tools for diagnosing the *laser* performance, for studying pulse *propagation* and laser-matter *interaction* processes as well as for encoding and decoding of *spatial phase information* in optical processors and communication systems would be available.

Two basic measuring situations for polychromatic phase fronts are schematically illustrated in Fig. 186a and b. The first picture on the left hand side (Fig. 186a) shows three different (time averaged) wavefronts at three different wavelengths. A conventional Shack-Hartmann wavefront sensor averages over the spectral wavefront information. The second picture (Fig. 186b) illustrates the option to spatially address the spectral phase front for a high-speed parallel data transfer. The typically very complex spectra of highly intense few-cycle pulses are challenging for pulse

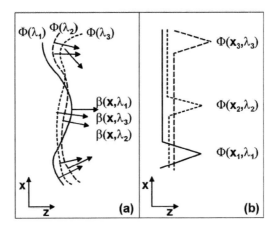

Fig. 186. Basic situations for the time-integrated measurement of space-variant spectral phase fronts: (a) spectral dependence of the local wavefront tilt (represented by local Poynting vectors for each partial spectral wavefront), (b) spatially encoded spectral phase front information.

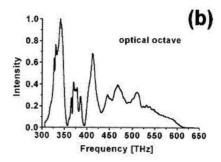

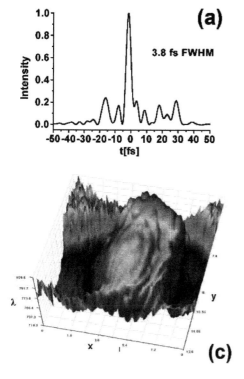

(c)

Fig. 187. Challenges to the pulse characterization at extreme parameters:

(a) sub-4-fs Ti:sapphire laser pulse generated by broadening the spectrum of an amplified pulse by self-phase modulation in gas-filled hollow waveguides and re-compressing by advanced types of chirped mirrors,

(b) ultrabroad (octave-spanning) spectral intensity distribution,

(c) spatial chirp of the output signal of a Ti:sapphire laser oscillator after propagating through an optical relay system.

characterization as well as read-out of spatio-temporal phase information, see the examples in Fig. 187a,b and c.

Recently, other groups compared spectrally selected and spectrally integrated wavefront measurements with a Shack-Hartmann-sensor at 5-fs pulse duration [2005_23]. The different spectral contributions were subsequently selected with a *set of spectral filters*. The difference between the wavefront reconstructed from spectrally resolved data and the averaged measurement was moderate. However, it is expected that significant discrepancies will be obtained at shorter pulse durations and higher intensities which are not being tolerable.

Digital holography with a *tunable filter* was also applied to reconstruct the polychromatic phase distribution [2004_10]. This method is limited by the sequential operation mode as well. It requires the separation of a very defined monochromatic reference wave so that the method is restricted to low intensities. The principal limits of this interesting technique, however, are not exactly defined so that it is worth to be a subject of further investigations.

The optimum solution for a polychromatic detection of local phase fronts would be to implement *hyperspectral detection systems* in a spatially resolving wavefront sensor or autocorrelator. Hyperspectral (multispectral) devices are characterized by combining the functionalities of spatial and spectral resolution with each other. In contrast to *sequential approaches* like multispectral cameras containing rotating or tunable filters, the main advantages of *parallel processing hyperspectral systems* are to reduce the measuring time and to be capable for real-time measurements (except for collinear autocorrelation or similar procedures with step-wise tuned delay lines). For a system of N parallel channels, the resulting multichannel detection time constant T_{MC} is roughly proportional to the inverse number of channels:

$$T_{MC} \propto \frac{1}{N} T_{SC} \tag{132}$$

(T_{SC} - single channel detection time constant). Depending on the number of unit cells of the array and the processing speed and architecture of electronic processors, additional time constants may have to be taken into account.

To realize the goal of parallel hyperspectral processing, different basic concepts are possible:

(a) spatial separation of spectral contributions in spectrally selective beam array generators containing DOE or programmable diffractive devices based on SLM,
(b) detectors with a transversal selection of spectral contributions by pixel material or filter arrays,
(c) multilevel detectors consisting of spectrally selective layers

As it was mentioned in Chapter 2, the application of arrays of hybrid refractive-diffractive microlenses and DOE to aberration correction and hyperspectral sensing and imaging tasks was reported in the literature (see, e.g., [1994_1]). Spectrally selective beam array generators or arrayed detectors with a local multichannel spectral resolution, however, were not applied to polychromatic pulses up to now.

Very recently, the application of thin-film *graxicons* (i.e. the combination of gratings and axicons, see paragraph 4.7.2.) to the hyperspectral mapping of ultrafast pulses was proposed [P16]. The results of experimental proof of this principle will be presented now.

7.4.2. Hyperspectral Shack-Hartmann wavefront sensor with graxicon arrays

The schematic setup of a hyperspectral axicon-based Shack-Hartmann sensor (HASHS) is drawn in Fig. 188 [2006_8]. In this setup, the reflective microaxicons of an ASHS are replaced by an array of hybrid reflective-diffractive elements we referred to as "graxicons" (see Figs. 75 and 76). In first test experiments, an Al grating with a period of 1.5 µm and a groove depth about 300 nm was covered with transparent thin-film microaxicons (hexagonal array of Gaussian-shaped fused silica). The layer thickness was < 1 µm and the period 405 µm. The light passes the fused silica layer twice and is transformed into an array of sub-beams similar to the case of pseudo-reflective compound structures (transparent layer on metal mirror, see Figs. 72 and 73). At polychromatic illumination, however, the sub-beams are needle beams only in one dimension but spectrally dispersed in the other one (orthogonal to the first). Shape and intensity envelope of the separated spectra contain information on the local spectral phase. In the experiment, the spectrally dispersed spots were magnified by a microscope objective and imaged by a zoom lens onto a high-resolution, high-dynamic-range CCD camera (Vosskuehler, 4 Megapixels).

Ultrashort pulses from a 10-fs Ti:sapphire laser oscillator (center wavelength 790 nm) were propagated through the hyperspectral system to prove the capability of the graxicon array for shaping an array of spectrally resolved focal zones. The incident angle at the graxicons was chosen to be about 26°. Figures 189 and 190 show a measured 2D spectral map and a spatio-spectral intensity profile, respectively. The spectral map was calibrated by a scanning fiber-based spectrometer (spatial resolution by an implemented pinhole: 15 µm, step width: 20 µm). The distributions indicate that the HASHS delivers spatial information on the spectral partitions of the beam. The system was not optimized with respect to focus diameter and grating parameters. For extraction of the spatio-spectral phase, the calibration has to be further improved (e.g. by reference waves, periodical filters, combs of absorption or emission lines).

An extension towards a spatio-spectral holographic autocorrelation by using a reference wave appears to be a further option to exploit graxicons for diagnostics of complex wavepackets at extreme pulse parameters.

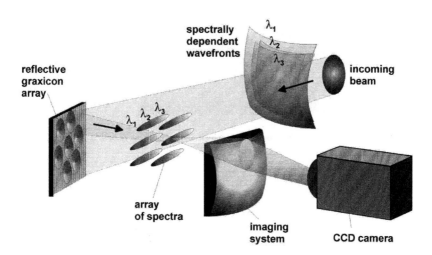

Fig. 188. Setup of a hyperspectral Shack-Hartmann sensor for time-averaged spectral phase front detection (schematically). An array of spectra is generated by a matrix of graxicons (refractive-diffractive elements combining the properties of gratings and thin-film microaxicons).

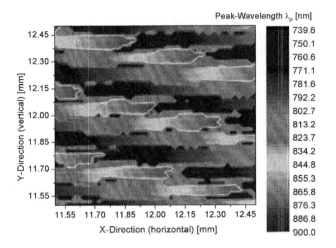

Fig. 189. Spectral map of a 10-fs Ti:sapphire laser pulse obtained with a hyperspectral Shack-Hartmann sensor (HSHS) with a graxicon array. The intensity pattern was calibrated with a fiber-based spectrometer (spatial resolution 15 μm).

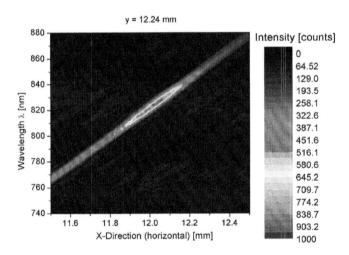

Fig. 190. Spatio-spectral intensity profile in x-direction for a selected focal zone (fixed y-position at 12.24 mm) (grayscale code: normalized intensity).

7.5. Generation of optical spatio-temporal X-pulses with thin-film structures

7.5.1. X-waves and X-pulses as spectral interference phenomena in spatio-temporal domain

The generation of Bessel-like pseudo-nondiffracting beams from *converging conical waves* of particular angular spectra was discussed in extenso in Chapter 6. At ultrashort pulse durations, specific spatio-temporal modifications appear because of spectral interference at extremely *broad bandwidths* [1999_13].

In early theoretical studies, packet-like solutions of the Maxwell equations were found by Brittingham (by him referred to as focus wave modes) [1983_3]. In generalized descriptions, classes of packet-like solutions of the homogeneous wave equation in free space were predicted by Belanger [1984_7] and Ziolkowski [1985_6]. In their classical paper on nondiffracting acoustical X-waves, Lu and Greenleaf found that bullet-like wave phenomena of a characteristic X-shape in space and time coordinates occur if broadband pulses are shaped in conical angular distributions. After the theoretical prediction in 1992 [1992_2], the effect could be experimentally verified [1992_3] and was first of practical impact for ultrasound transducers and later for shaping radio waves and THz radiation [1999_15, 2000_12]. Furthermore, axial uniformization effects were pointed out for polychromatic Bessel beams [1996_9]. Recently it was argued that broadband surface ocean waves of certain angular spectra can cause the so-called *monster wave* phenomena in analogous way to X-waves [2004_08]. In this picture, pulsed X-waves or X-pulses can be regarded to be optical monster waves - reaching giant amplitudes and concentrating most of their energy in a small volume of time-space continuum. This is but a consequence of one more free parameter compared to the monochromatic (stationary) case and agrees with the known statement of information theory that the *degree of localization* can be improved by enhancing the spectral bandwidth.

Optical X-waves have first been demonstrated in pioneering cross correlation experiments of the group of Peeter Saari with high-pressure lamps continuously emitting at coherence lengths in fs-range [1997_40]. Pulsed X-waves or X-pulses were indirectly proved in experiments with a pulsed Ti:sapphire laser at pulse durations about 210 fs [1997_26]. As it could be shown later [2004_21], the coherence time was too long in this case to obtain a high contrast X-shape directly in spatio-temporal experiments.

With the access to ultrashort-pulse sources and computing capability for an extended simulation of linear and nonlinear-optical propagation, an ever increasing interest in pulsed X-waves [1998_12], pulsed Bessel beams [1998_13], and related phenomena [2004_18] could be noticed. The spatially induced group velocity dispersion of ultrashort-pulsed Bessel beams was analyzed [2002_12]. Methods for a representation of X-waves by their angular spectrum [1996_8] and their space-frequency analysis [1997_15] were developed. Generalizations including special cases like orthogonal X-waves [2001_19], Y-waves [2000_19] and interfering localized waves and "frozen waves" [2005_15] were proposed.

In theory and experiments it was confirmed that localized waves indeed can be optically realized [2002_43, 2004_31, 2006_13]. Advanced theoretical investigations on relativistic Doppler effect [2003_30] and Gouy effect [1989_9, 2003_30] might initiate corresponding experiments in near future. Recently, even the possibility to overcome the photon localization barrier was discussed [2005_22, 2006_10]. Apart from the basic research, the interest in X-pulses was also driven by the enormous application potential of localized wavepackets in space and time, e.g. for ultrafast materials processing [2001_37] or optical communication (as originally proposed by Lu and He [1999_17]. One of the most controversially discussed features of localized waves, their superluminality [2000_10, 2003_13], could be successfully verified by ionization experiments [2002_27] and might have impact for high-speed laser-triggered switching or the excitation of photochemical reactions at superluminal velocity.

The propagation and interaction of localized waves in dispersive media [1996_27] and the generation of so-called "light-bullets" (i.e. soliton-like localized solutions under nonlinear conditions) is subject of numerous investigations [1997_14, 2003_4, 2003_21, 2005_17, 2006_14, 2006_15] and stimulates the research for ultrafast communication and metrology through fibers, waveguides and other dispersive or otherwise lossy media.

The *first direct detection of pulsed free-space Bessel-like beams* was enabled by the application of small-angle thin-film axicons. Two types of X-pulses and modified X-pulses were generated with extended conical thin-film axicon mirrors as well as Gaussian-shaped thin-film microaxicons [2002_7, 2003_11, 2004_08, 2004_9]. These results will briefly be discussed in the following paragraph.

7.5.2. Generation and direct detection of arrayed microscopic-size pulsed optical Bessel-like X-waves and single macroscopic Bessel-Gauss X-pulses

As already mentioned in Chapter 6, the generation of *arrays of ultrashort-pulsed Bessel-like microscopic-size beams* was first performed with thin-film microaxicons [2000_4, 2000_6, 2001_8, 2002_5]. However, the characteristic features of *Bessel-like X-waves* could not be identified in these initial experiments because of the lack of an appropriate technique to analyze their spatio-temporal structure with high resolution and low distortion. The *first direct experimental indication* of the spatio-temporal X-shape of Bessel-like pulses was enabled by an experimental setup (schematically in

Fig. 191). Here, a balanced interferometric (collinear) autocorrelator of Mach-Zehnder architecture was applied in contrast to the matrix autocorrelators shown before. The essential progress of the arrangement consisted in matching the following necessary conditions:

- The availability of low-dispersion microoptical array beam shapers based on thin-film microaxicons of Gaussian shape with extremely small conical angles.
- An appropriate detection method of appropriate spatial and temporal resolution working with separated channels without cross-talk.

Thus it was possible to generate and to detect X-pulses generated from an amplified and re-compressed Ti:sapphire fs-laser system at pulse durations between 7 and 25 fs [2001_10, 2002_7, 2002_9, 2003_11, 2004_21]. The experiment was the first application of two-dimensional wavefront autocorrelation to arrays of such short pulses. A two-dimensional cut through the 3D spatio-temporal autocorrelation function is plotted in Fig. 192 [2003_11, 2004_8, 2004_9, 2006_5]. Because of the small diameter of the generating elements (Gaussian waist radius 150 μm), the illuminating light distribution per element can be approximated by a plane wave. The spatio-temporal characteristics was found to well agree with theoretical simulations of the

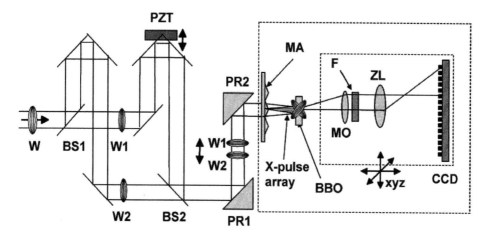

Fig. 191. Experimental setup with a balanced spatio-temporal interferometric autocorrelator for the direct detection of optical Bessel-like X-pulses generated from Ti:sapphire laser pulses at pulse durations of 7-25 fs (W - incoming wavepacket, BS1 and BS2 - beam splitters, W1 and W2 - replica of the wavepacket, PZT - piezo translator, PR1 and PR2 - prism reflectors at small incident angles, MA - thin-film microaxicon array with Gaussian-shaped microaxicons, BBO - 10-100 μm thick NLO crystals for second harmonic generation, MO - microscope objective, F - broadband metal filter, ZL - zoom lens, CCD - matrix camera of 1300 x 1020 pixels).
Figure reprinted with permission from Phys. Rev. A 67, 063820 (2003) [2003_7]. Copyright 2003 by the American Physical Society.

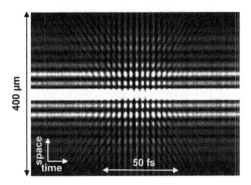

Fig. 192. Two-dimensional projection of a three-dimensional data set for the spatio-temporal autocorrelation function of a selected sub-10-fs wavepacket (part of an array of Bessel-like pulses generated in free space by a matrix of fused silica Gaussian-shaped thin-film elements deposited on silica, spatial period 405 μm). The interference pattern in space and time clearly reveals the characteristic shape of an X-pulse.

polychromatic beam propagation [2004_21]. Arrays of up to > 1000 Bessel-like sub-beams were generated and analyzed with the wavefront autocorrelation method. The self-reconstruction properties, angular aberrations, truncation, focusing behavior and interaction with materials (glass, quartz) were investigated. With focused Bessel-like beams in media, pulse shortening effects were found. Wavepackets with pulse durations down to the single cycle range were indicated. The detailed mechanism for this phenomenon, however, is currently still under investigation.

Pulsed Bessel-Gauss beams were obtained by illuminating large-size ultraflat thin-film structures of 1 cm diameter (fabricated at Laval University, Quebec) [2002_18]. By working at small angles of incidence, spherical aberrations could be minimized in reflective arrangements. At large incident angles, Mathieu-like beams and arrays of Mathieu-like beams were generated (compare Fig. 182).

7.6. Self-apodized truncation of ultrashort and ultrabroadband Bessel pulses

The principle of *self-apodized truncation* was already discussed in detail (see paragraph 6.3.5.). Here we like to address the essentially different properties of beams produced by this spatial beam shaping method compared to Gaussian beams [2004_11]. As it will be shown, the spatio-spectral and spatio-temporal properties of truncated Bessel-like wavepackets propagating along extremely extended, needle-shaped volumina are of *particular relevance for ultrashort, ultrabroadband pulses*.

7.6.1. Spatial propagation of ultrashort-pulsed and ultrabroadband truncated Bessel-like beams generated with thin-film beam shapers

The spatial propagation of Bessel-like pulses truncated in self-apodized regime (i.e. approximately without diffraction at the edge of the aperture) was theoretically and experimentally studied [2004_11, 2005_7. 2005_8, 2005_12, 2006_5]. Experiments were performed with a Ti:sapphire laser system (pulse durations 5-30 fs, spectral FWHM 40-240 nm). Small-angle *Bessel-Gauss pulses* of an about 1 m deep zone of non-spreading propagation were shaped by illuminating a flat concave, circularly symmetric Au axicon of 1 cm diameter (conical angle 0.05° [2002_18]) with a nearly

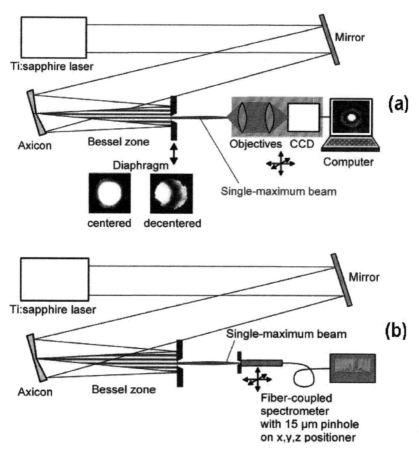

Fig. 193. Experimental setups for self-apodized truncation of Bessel-Gauss pulses: Setup for (a) adjusting the truncating diaphragm and monitoring the propagation in axial direction, (b) spatial mapping of the pulse spectrum by a fiber-based spectrometer with a pinhole at the end face.

Gaussian shaped beam profile. The input beam diameter was adapted by a reflective beam expander. The axicon was illuminated at incident angles $< 20°$. Intensity distributions at different propagation distances were imaged onto a CCD camera (1300 x 1030 pixels, zoom objective 1:1.2/12.5-75 mm, microscope M = 4x magnification). With the help of the monitored image and by axially translating the aperture in order to exactly match the fringes it was possible to accurately adjust the truncation setup (schematically in Fig. 193a). Spatio-spectral maps were detected by scanning the planes of interest with a fiber-based spectrometer (Ocean Optics) (schematically in Fig. 193b). The spatio-temporal transfer was investigated with a commercial (spatially integrating) autocorrelator (APE).

The spatial resolution was defined by a small diaphragm (15-μm diameter). The truncating aperture (diameter about 500 μm) was placed at the beginning of the zone of maximum intensity (about 110 cm distance from the axicon). The intensity propagation behind the aperture is plotted in Fig. 194.

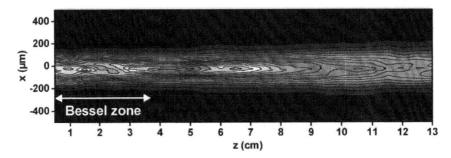

Fig. 194. Intensity propagation of a truncated Bessel-Gauss beam generated from 10-fs Ti:sapphire laser pulses with a thin-film axicon (1 cm diameter, 0.05° conical angle). The Rayleigh length was found to be > 10 cm. A region in near field (here referred to as "Bessel zone") appears without any beam spreading. (Please notice the different scales in z- and x-directions.)

A Rayleigh length of 13 cm was detected. The corresponding aspect ratio (Rayleigh length : initial waist radius) was found to be about 440 : 1. At a pulse duration of < 5 fs, an even higher value of 520 : 1 was obtained. Furthermore, a *near field* region with obviously no beam spreading was indicated (> 3.5 cm depth). Because of the nondiffracting characteristics of this zone it was referred to as *"Bessel zone"* (in analogy to the zone of undistorted progressive beam propagation of pseudo-nondiffracting beams). This result agrees with theoretical calculations based on Rayleigh-Sommerfeld diffraction theory at polychromatic illumination for realistic spectra. In Fig. 195, the simulated propagation for Bessel-Gauss beams generated by a conical axicon mirror is compared to the beam propagation of a focused Gaussian beam. The beam parameters for both types of beams were chosen to deliver the same

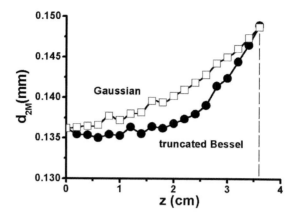

Fig. 195. Comparison between a polychromatic truncated Bessel-Gauss pulse and a focused polychromatic Gaussian beam in the near field for identical spectra (corresponding to a pulse duration of 10 fs and a spectral FWHM of 120 nm) and initial diameter. The curves show the second-order diameters of both as a function of the propagation path in axial direction. The dashed line marks the crossing points of Gaussian and truncated Bessel distributions

radial second moment diameter d_{2M} in the initial planes (aperture plane of truncated Bessel-Gauss beam and focus of Gaussian beam).

In the far field, the Gaussian beam has a slightly smaller diameter (as it was expected). In the near field, however, the diameter of the truncated Bessel beam is smaller. This results from starting with a radial distribution function different to the Gaussian one (central maximum of the squared Bessel function). It can be concluded that a structured beam shaping with thin-film components enables for tailoring non-spreading near field zones over distances even in *few-centimeter range*.

7.6.2. Spatio-spectral and spatio-temporal transfer of ultrashort-pulsed and ultrabroadband truncated Bessel-Gauss beams

Intensity transfer functions for compact description of a system response to an ultrashort pulse can be defined for the time domain and for the spectral domain. In the time domain, the output signal $I'(t)$ is the convolution of signal $I(t)$ and temporal transfer function T_i:

$$I'(t) = \int_{0}^{\infty} T_i(\tau) I(t - \tau) \, d\tau = T_i(t) \otimes I(t) \ . \tag{133}$$

In the spectral domain, the output signal $I'(\omega)$ represents a multiplication of signal spectrum $I(\omega)$ and spectral transfer function $S_i(\omega)$:

$$I'(\omega) = S_i(\omega) \cdot I(\omega) \ . \tag{134}$$

For bandwidth-limited pulses, temporal and spectral transfer functions are related to each other by Fourier transform. Assuming dependence of the spectrum on the spatial coordinates (x,y), a *spatio-spectral intensity transfer function* can be defined as

$$I'(x, y, \omega) = S_i(x, y, \omega) \cdot I(x, y, \omega) \ . \tag{135}$$

The *spatio-spectral transfer functions* of Bessel beams were detected with the scanning setup as shown in Fig. 193b. The spectral distributions were analyzed on the basis of statistical moments up to the 4th moment [2006_22]. It was found that the fringes of Bessel-Gauss pulses are *significantly modulated* with respect to their

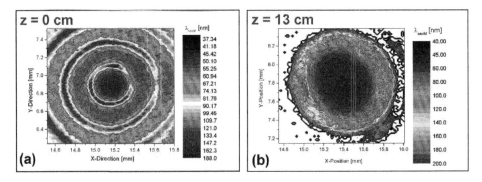

Fig. 196. Measured spatio-spectral maps for the rms standard deviation of the spectra directly behind the truncating aperture and at a distance of 13 cm.

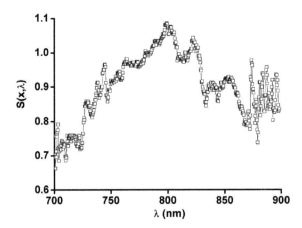

Fig. 197. Spectral transfer function for one position (corresponding to the propagation of the truncated Bessel-Gauss pulse in Fig. 194 and spectral data in Fig. 196).

bandwidth as well as particular features of their spectrum (e.g. *screwness and kurtosis*). If the central lobe is truncated in self-apodizing mode (see Chapter 6), however, the distribution propagates over more than a complete Rayleigh length (here 10-13 cm depending on the axial position of the diaphragm) with relative small change in the spectral bandwidth of the beam center (Figs. 196a and b). The corresponding relative spectral transfer function for an arbitrary position in the beam center is shown in Fig. 197 (values > 1 result from the influence of fluctuations on the mathematical analysis). For nearly the FWHM of the spectrum (120 nm), the transfer factor was found to be > 0.9. The results lead to the conclusion that the *pseudo-nondiffracting nature of such beams is also indicated by their spatio-spectral characteristics* (spectrally invariant propagation). This is also supported by temporal autocorrelation measurements at distances of 0 and 13 cm from the truncating aperture (measured pulse durations of 10.5 and 10.3 fs, respectively). Because of the internal optical paths within the autocorrelator, the real distances were even larger than in the case of the spectral mapping but in fact the temporal behavior appears to be unaffected.

7.6.3. Comparison to ultrashort-pulsed Gaussian beams

The Rayleigh length $z_{1/2}$ of ideal, monochromatic *Gaussian beams* depends on the inverse of the wavelength λ and square of the Gaussian waist radius w_0:

$$z_{1/2}(\lambda) = \frac{\pi \cdot w_0^2(\lambda)}{\lambda} \ . \tag{136}$$

In the case of a *polychromatic* beam (ultrashort pulse) this leads necessarily to a spatially inhomogeneous spectral distribution depending on the initial conditions (spatio-spectral distribution in the input plane [1996_33, 1997_48, 1998_19]) even if spectral and travel-time related dispersion effects are neglected.

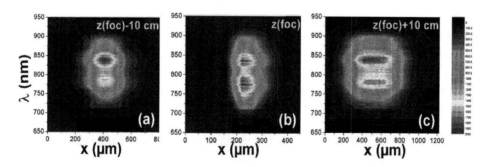

Fig. 198. Spatio-spectral maps for a propagating Gaussian beam of a Rayleigh length comparable to that of the truncated Bessel beam.

An extended Gaussian focal zone of *comparable Rayleigh length* (> 10 cm) to the truncated Bessel beam was experimentally generated by illuminating a concave gold mirror (radius of curvature 5m) at a small angle of incidence [2004_11]. Radial cuts through intensity patterns at different axial positions in steps of 10 cm are drawn in Figs. 198a-c. It can be recognized that the spectral distribution in one spatial direction is not constant along the axis.

From the spectral transfer measurements with ultrashort-pulsed Bessel-Gauss and Gaussian beams it can be concluded that the main advantage of truncated Bessel-like beams are

- to be nondiffracting with respect to their spatio-spectral characteristics as well as their spatio-temporal characteristics,
- to generate extended focal zones at much shorter distances so that pulse broadening in air can be avoided and more compact systems can be designed.

The drawback of the loss in the outer fringes could either be compensated by recycling the fringe energy by coherent addition [P13] or shaping needle beams of larger diameters. This topic is still under investigation.

7.7. Spatio-temporal self-reconstruction and nondiffracting images

7.7.1. Self-reconstruction and spatio-temporal information

The spatial self-reconstruction properties of Bessel-like beams were already addressed in the previous Chapter. The simple picture of a shadow zone holds only for monochromatic beams whereas in the case of polychromatic and ultrafast-pulsed fields (*spatio-temporal self-reconstruction*), the system response is more complicated. The reasons for that are diffraction, spectral dispersion in media, nonlinear effects at the typically high intensities and spatio-temporal coupling [2000_33].

Recently it was demonstrated that a specific information (e.g. on the X-wave structure of a pulsed Bessel-like beam) can be reconstructed even at significant perturbations (for more details, see [2004_12, 2006_05]). The high robustness and spatial localization of pseudo-nondiffracting beams can be exploited to transfer data and reconstruct information in measuring systems. The spectral bandwidth and the high temporal resolution of ultrashort pulses enables for further applications by additional degrees of freedom. A communication via addressed X-pulses [1999_17] and networks with solitons [2005_25] were proposed in the literature. Multipixel light sources were developed as a step in this direction [2003_20].

The combination of the potential of massive parallel information transfer via complete spatio-temporal and spatio-spectral images (Saari's "flying image" [1996_28]) and the properties of undistorted propagation and self-reconstruction of pseudo-nondiffracting beams is currently on the roadmap.

7.7.2. Nondiffracting images

To illustrate this exciting and dynamic field of research by a selected example, the generation of ultrashort-pulse self-reconstructing nondiffracting images will be shown. The concept consists in combining addressable spatial light modulators (to spatially encode amplitude and phase information in a wavepacket) with the nondiffracting and self-reconstructing behavior of arrays of Bessel-like beams (Fig. 199) [2004_12].

In first experiments, a setup as depicted in Fig. 200 was applied. An amplitude/phase information was programmed into a liquid-crystal-on-silicon SLM (LCoS-SLM, Holoeye Photonics, LC-R 3000, 1920x1200 pixels). The beam of a 10-fs Ti:sapphire laser (center wavelength 790 nm, spectral FWHM 120 nm) was reflected at the SLM and passed a transparent thin-film microaxicon (Gaussian-shaped fused silica microaxicons, period 405 μm, structure height 3 μm). The binary pattern from SLM (here the capital letter "E") was transformed in an array of Bessel-like X-pulses propagating over the axial extension of the Bessel zone as nondiffracting, self-reconstructing beamlets with broadband spectral transfer functions. The "flying image" was amplitude-decoded with an IR polarizer and the resulting intensity pattern imaged onto a high-resolution high-dynamic range CCD-camera (Vosskuehler, 4 million pixels, 12 bit). Experimental results are presented in Fig. 201 [2006_5]. In the three pictures

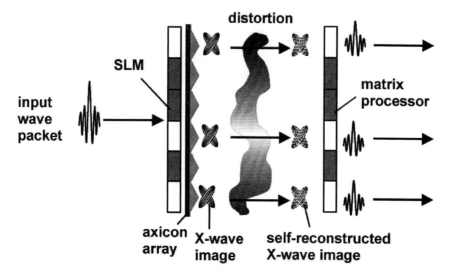

Fig. 199. Concept of nondiffracting images: combination of addressable SLM and nondiffracting beam arrays generated by low-dispersion thin-film microaxicons (schematically).

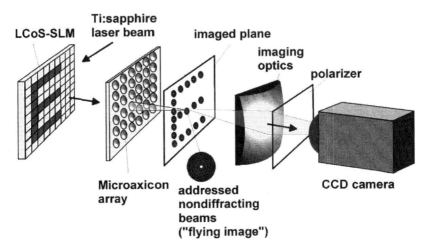

Fig. 200. Experimental setup for nondiffracting ultrafast image experiments with a combination of an LCoS-SLM and an array of microaxicons (schematically).

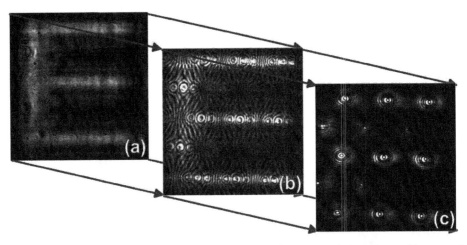

Fig. 201. Nondiffracting ultrafast image: (a) undistorted image signal from SLM, (b) array of Bessel-like beams significantly distorted by inserting an amplitude grating, (c) array of partially self-reconstructed Bessel-like beams carrying spatial and temporal information (Bessel beam fringes, X-pulse structure) enabling to reconstruct image information.

one can see how the initial intensity pattern looks like (a), how the array of Bessel-like sub-beams are strongly distorted by an amplitude grating (b), and how finally an array of partially self-reconstructed sub-beams reveals the encoded image information again (c). The decoding of spatio-temporal image information by X-pulse recognition will be a subject of continuing activities including wavefront autocorrelators.

Chapter 8

OUTLOOK

In this book, thin-film microoptics was introduced as a dynamic and promising field which combines the specific features of small-scale optics with those of thin optical films to a new and specific functionality. At this point it should be clearly pointed out that the goal of thin-film microoptics was not simply to replace conventional microoptical elements by their thin-film equivalent. On the contrary, high-precision microoptical systems are indeed matching the requirements of numerous applications in an optimum way. The special charm of thin-film microoptics is rather to *extend* the capabilities of microoptical systems in very outer regions with respect to the parameters of optical fields to be transformed. Their unusual and sometimes even surprising properties arise from low dispersion, spectral degrees of freedom, nonspherical shapes, small-angle design, mechanical flexibility and/or further particular features which can be well realized with thin-film approaches.

For the spatio-temporal *shaping and diagnostics of ultrashort wavepackets*, new prospects were opened by novel matrix processor architectures based on small-angle axicon arrays. Such components enable for a generation of *Bessel-like and needle-shaped nondiffracting beams* of unique propagation properties (e.g. self-reconstruction or robustness against tilt). It was possible to generate and directly detect localized few-cycle wave-phenomena (*X-pulses*) for the first time with full resolution in space-time-coordinates [2002_7, 2002_9]. By combining the well-known principles of autocorrelator and Shack-Hartmann sensor in *wavefront autocorrelation* experiments, a *more complete* characterization of ultrashort pulses and interactions with respect to the spatial domain was obtained. It was shown that image information can be encoded in arrays of nondiffracting micro-beams [2006_6].

First steps to a *hyperspectral* diagnosis of the spatio-spectral phase with graxicons were reported [2006_8]. The exploitation of the spectral degrees of freedom for tailoring pulsed wave fields and processing optical information is only in the beginning.

One of the great challenges in designing next generation microoptical devices will be to fuse the advantageous features of both spatial and temporal beam shaping in more universal *beam synthesizers* which should be *addressable in space and time characteristics* (say, in a way like a printer head). New approximations of the old vision of programmable phase and amplitude shaping systems [1992_7] will be invented.

The scale of thin-film microoptical applications reaches from creating extremely extended focal zones to very small features on the *nanoscale*. With access to nanolayer microoptics, a new world of refraction-dominated beam shaping at the limits of *ultraflatness* has to be explored. The integration of layers consisting of *nonlinear* or fluorescent nanocrystalline materials will lead to advanced functionality as well. Recent *nanostructuring* experiments in other groups demonstrated that shadow-mask deposition methods are far from being at the end of their potential [2004_38, 2005_30, 2005_31]. The combination of classical lithographic techniques with thin-film microoptical structuring is currently under investigation [2006_9]. Finally, thin films with properties of negative index materials [2000_8] would surely be exciting modules for future optical construction kits.

The author hopes to have convincingly shown that in the "flatland" of thin microoptical films there is space enough for many more innovative approaches. The list of unsolved problems is not short. But we know for sure:

There always exists one more solution than you initially thought...

REFERENCES

[1836_1] W. H. F. Talbot, Facts relating to optical sciences. No. IV., Philos. Mag. **9** (Third series), 401-407 (1836).

[1884_1] E. Abott Abott, *Flatland - A romance of many dimensions*, Second revised edition, 1884; see: http://www.ibiblio.org/eldritch/eaa/fl.htm.

[1890_1] C. R. Gouy, Sur une propriete nouvelles des ondes lumineuses, Comp. Rendue Acad. Sci. Paris **110**, 1251 (1890).

[1890_2] C. R. Gouy, Sur la propagation anomale des ondes, Comp. Rendue Acad. Sci. Paris **111**, 33 (1890).

[1917_1] A. Einstein, Zur Quantentheorie der Strahlung, Phys. ZS **18**, 121-128 (1917).

[1941_1] J. A. Stratton, *Electromagnetic theory*, McGraw Hill, New York (1941).

[1947_1] A. Herpin, Calcul du pouvoir réflecteur d'un système stratifiè quelconque, C. R. Acad. Sci. **225**, 182-183 (1947).

[1950_1] F. Abelès, Recherches sur la propagation des ondes électromagnetiques sinusoidales dans les milieus stratifiés, Ann. Phys. Ser. **12**, 706-784 (1950).

[1952_1] L. Holland and Steckelmacher, The distribution of thin films condensed on surfaces by the vacuum evaporation method, Vacuum **2**, 346-364 (1952).

[1954_1] J. H. McLeod, The axicon: A new type of optical element, J. Opt. Soc. Am. **44**, 592-597 (1954).

[1957_1] F. A. Jenkins and H. E. White, *Fundamentals of optics*, McGraw-Hill Book Co., New York, 1957.

[1962_1] T. Asakura, Diffraction patterns by non-uniform phase and amplitude aperture illumination (I), Oyo Butsuri (J. Appl. Phys. Japan) **31**, 730 (1962) (in Japanese).

[1962_2] R. Barakat, Solution of the Luneburg apodization problem, J. Opt. Soc. Am. A **52**, 264-275 (1962).

[1964_1] F. Gires and P. Tournois, Interferometre utilisable pour la compression d'impulsions lumineuses modulees en frequence, C. R. Acad. Sci. Paris **258**, 6112-6115 (1964).

[1964_2] P. Jacquinot and B. Roizen-Dossier, Apodisation, in: E. Wolf (Ed.), *Progr. in Optics*, Vol. **3**, Elsevier, 1964, p. 29.

[1964_3] P. Giacomo, B. Roizen-Dossier, and S. Roizen, Preparation par evaporation sous vide d'apodiseur circulaires, J. de Phys. (France), **25**, 285-290 (1964).

[1965_1] J. T. Winthrop and C. R. Worthington, Theory of Fresnel images. I. Plane periodic objects in monochromatic light, J. Opt. Soc. Am. **55**, 373-381 (1965).

[1965_2] T. Asakura and Y. Kikuchi, Diffraction patterns by non-uniform phase and amplitude aperture illumination, V. Diffraction experiment with amplitude filter, Oyo Butsuri (J. Appl. Phys. Japan) **34**, 795-801 (1965).

[1967_1] W. D. Montgomery, Self-imaging objects of infinite aperture, J. Opt. Soc. Am. **57**, 772-778 (1967).

[1968_1] W. D. Montgomery, Algebraic formulation of diffraction applied to self-imaging, J. Opt. Soc. Am. **58**, 1112-1124 (1968).

[1968_2] R. P. Howson, Optical properties of thin films of indium arsenide, Brit. J. Appl. Phys. (J. Phys. D), **1**, 15-23 (1968).

[1970_1] A. Musset and A. Thelen, Multilayer antireflection coatings, in: E. Wolf (Ed.), *Progress in Optics*, Vol. VIII, North-Holland Publishing Co., Amsterdam, 1970.

[1970_2] A. A. Hoag and D. J. Schroeder, "Nonobjective" grating spectroscopy, PASP **82**, 1141-1145 (1970).

[1971_1] H. Dammann and K. Görtler, High-efficiency in-line multiple imaging by means of multiple phase holograms, Opt. Commun. **3**, 312-315 (1971).

[1971_2] R. Ulrich and R. J. Martin, Geometrical optics in thin film light guides, Appl. Opt. **10**, 2077-2085 (1971).

[1971_3] D. A. Palmer, A monochromator using graded interference filters, J. Phys. E: Sci. Instrum. **4**, 41-42 (1971).

[1972_1] L. d'Auria, J. P. Huignard, A. M. Roy, and E. Spitz, Photolithographic fabrication of thin film lenses, Opt. Commun. **5**, 233-235 (1972).

[1972_2] B. Ramprasad, T. Radha, and M. Rao, On uniformity of thin film thickness on rotating substrates, J. Vac. Sci. Technol. **9**, 1227-1231 (1972).

[1972_3] P. Bousquet, A. Fournier, R. Kowalczyk, E. Pelletier, and P. Roche, Optical filters: monitoring process allowing the auto-correction of thickness errors, Thin Solid Films **13**, 285-290 (1972).

[1972_4] H. A. Macleod, Turning value monitoring of narrow-band all-dielectric thin-film optical filters, Opt. Acta **19**, 1-28 (1972).

[1974_1] A. Papoulis, Ambiguity function in Fourier optics, J. Opt. Soc. Am. **64**, 7709-788 (1974).

[1974_2] G. L. McAllister, W. H. Steier, and W. B. Lacina, Improved mode properties of unstable resonators with tapered reflectivity mirrors and shaped apertures, IEEE J. Quant. Electron. **10**, 346-355 (1974).

[1974_3] L. P. Boivin, Thin-film laser-to-fiber coupler, Appl. Opt. **13**, 391-395 (1974).

[1974_4] A. J. Campillo, B. Carpenter, B. E. Newman, and S. L. Shapiro, Soft apertures for reducing damage in high-power laser amplifier systems, Opt. Commun. **10**, 313-315 (1974).

[1974_5] H. Slevogt, *Technische Optik*, Sammlung Göschen Bd. 9002, Walter de Gruyter, Berlin, 1974 (in German).

[1975_1] G. A. Horridge (Ed.), *The compound eye and vision of insects*, Clarendon Press, Oxford, 1975.

[1975_2] Lawrence Livermore National Laboratory, *Annual Report*, UCRL-50021-74, 1975.

[1976_1] I. K. Krasjuk, S. G. Lukishova, P. P. Pashinin, A. M. Prokhorov, and A. V. Shirkov, Formation of the spatial profile of a laser beam with the use of soft apertures, Soviet Phys.: Quant. Electr. **3**, 1337-1340 (1976) (in Russian).

[1977_1] W. Gellert, H. Kästner, and S. Neubert, *Lexikon der Mathematik*, VEB Bibliographisches Institut, Leipzig, 1977 (in German), 623.

[1978_1] S. K. Yao and D. B. Anderson, Shadow sputtered diffraction-limited waveguide Luneburg lenses, Appl. Phys. Lett. **33**, 307-309 (1978).

[1978_2] S. K. Yao, D. B. Anderson, R. R. August, B. R. Yourmans, and C. M. Oania, Guided-wave optical thin-film Luneburg lenses: fabrication technique and properties, Appl. Opt. **24**, 4067-4079 (1978).

[1978_3] J. A. Dobrowolski and D. Lowe, Optical thin film synthesis program based on the use of Fourier transforms, Appl. Opt. **29**, 3039-3050 (1978).

[1979_1] S. K. Yao, Theoretical model of thin-film deposition with shadow effect, J. Appl. Phys. **50**, 3390-3395 (1979).

[1980_1] D. T. Moore, Gradient-index optics: a review, Applied Optics, **19**, 1035-1038 (1980).

[1980_2] A. Dubik, J. Firak, and K. Jach, *Diffraction on hard and profiled apertures in high-power laser systems*, Inst. of Plasma Phys. & Laser Fusion of S. Kalisky, Report IFPILM, N 4/80 (1980), Warsaw, 24 (in Polish).

[1980_3] R. W. Boyd, Intuitive explanation of the phase anomaly of focused light beams, J. Opt. Soc. Am. **70**, 877 (1980).

[1980_4] T. W. Hänsch and B. Couillaud, Laser frequency stabilization by polarization spectroscopy of a reflecting reference cavity, Opt. Commun. **35**, 441-444 (1980).

[1981_1] E. Colombini, Design of thin-film Luneburg lenses for maximum focal length control, Appl. Opt. **20**, 3589-3593 (1981).

[1981_2] S. B. Arifzhanov, R. A. Ganeev, A. A. Gulamov, V. I. Redkorechev and T. Usmanov, High optical quality beam shaping in cascaded Nd laser, Soviet Phys.: Quant. Electr. **8**, 1246-1252 (1981) (in Russian).

[1982_1] J. P. Borgogno, B. Lazarides, and E. Pelletier, Automatic determination of the optical constants of inhomogeneous thin films, Appl. Opt. **21**, 4020-

4029 (1982).

[1983_1]   P. J. Sands, Classification scheme and nomenclature for refractive-index distributions, Appl. Opt. **22**, 430-431 (1983).

[1983_2]   N. McCarthy and P. Lavigne, Optical resonators with Gaussian reflectivity mirrors: misalignment sensitivity, Appl. Opt. **22**, 2704-2708 (1983).

[1983_3]   J. N. Brittingham, Focus wave modes in homogeneous Maxwell equations: Transverse electric mode, J. Appl. Phys. **54**, 1179-1181 (1983).

[1984_1]   K. Iga, Y. Kokubun, and M. Oikawa, *Fundamentals of microoptics - distributed-index, microlens, and stacked planar optics*, Academic Press, Tokyo, 1984.

[1984_2]   B. Tatian, Fitting refractive-index data with the Sellmeier dispersion formula, Appl. Opt. **23**, 4477-4485 (1984).

[1984_3]   S. Misawa, M. Oikawa, and K. Iga, Maximum and effective numerical apertures of a planar microlens, Appl. Opt. **23**, 1784-1786 (1984).

[1984_4]   H. Haferkorn: *Optik*, Deutscher Verlag der Wissenschaften, Berlin, 1984 (in German), 114.

[1984_5]   I. Awai and J. I. Ikenoue, Effect of film transition layers on the Abeles method, Appl. Opt. **23**, 1890-1896 (1984).

[1984_6]   N. McCarthy and P. Lavigne, Optical resonators with Gaussian reflectivity mirrors: output beam characteristics, Appl. Opt. **23**, 3845-3850 (1984).

[1984_7]   P. A. Belanger, Packetlike solutions of the homogeneous wave equation, J. Opt. Soc. Am. **1**, 723-724 (1984).

[1985_1]   N. F. Borrelli, D. L. Morse, R. H. Bellman, and W. L. Morgan, Photolytic technique for producing microlenses in photosensitive glass, Applied Optics **24**, 2520-2525 (1985).

[1985_2]   Y. Koike, H. Hidaka, and Y. Ohtsuka, Plastic axial gradient-index lens, Appl. Opt. **24**, 4321-4325 (1985).

[1985_3]   J. J. Cowan, The holographic honeycomb microlens, Proc. SPIE Vol. **523**, 251-259 (1985).

[1985_4]   R. M. A. Azzam, Simultaneous reflection and refraction of light without change of polarization by a single-layer-coated dielectric surface, Opt. Lett. **10**, 107-109 (1985).

[1985_5]   P. Lavigne, N. McCarthy, and J. G. Demers, Design and characterization of complementary gaussian reflectivity mirrors, Appl. Opt. **24**, 2581-2586 (1985).

[1985_6]   R. W. Ziolkowski, Exact solutions of the wave equation with complex source locations, J. Math. Phys. **26**, 861-863 (1985).

[1985_7]   J.-C. M. Diels, J. J. Fontaine, I. C. McMichael, and F. Simoni, Control and measurement of ultrashort pulse shapes (in amplitude and phase) with femtosecond accuracy, Appl. Opt. **24**, 1270-82 (1985).

[1986_1] S. Wang, J. Z. Wilcox, M. Jansen, and J. J. Yang, In-phase locking in diffraction-coupled phased array diode lasers, Appl. Phys. Lett. **48**, 1770-1772 (1986).

[1986_2] K. Jain, Advances in excimer laser lithography, Proc. SPIE Vol. **710**, 35-42 (1986).

[1986_3] A. A. Mak, V. M. Mitkin, and G. P. Petrovsky, Formation of gradient refractive index of glass by laser radiation, Dok. Akad. Sci. USSR **287**, 845-849 (1986).

[1986_4] V. P. Veiko, K. G. Predko, V. P. Volkov, and P. A. Skiba, Laser formation of micro-optical elements based on glass-ceramics materials, Dok. Akad. Sci. USSR **287**, 845-849 (1986).

[1986_5] F. P. Schäfer, On some properties of axicons, Appl. Phys. B **39**, 1-8 (1986).

[1987_1] H. Nishihara and T. Suhara, Micro Fresnel lenses, in: E. Wolf (Ed.), *Progress in Optics*, Vol. XXIV, North-Holland, Amsterdam, 1987, 1-40.

[1987_2] F. Gori, G. Guttari, and C. Padovani, Bessel-Gauss beams, Opt. Commun. **64**, 491-495 (1987).

[1987_3] M. A. Karim, A. K. Cherri, A. A. S. Awwal, and A. Basit, Refracting system for annular laser beam transformation, Appl. Opt. **26**, 2446-2449 (1987).

[1987_4] L. Hornak, Fresnel phase plate lenses for through-wafer optical interconnections, Appl. Opt. **26**, 3649-3654 (1987).

[1987_5] R. A. Haefer, *Oberflächen- und Dünnschicht-Technologie, Teil I, Beschichtungen von Oberflächen*, Springer, Berlin, 1987 (in German).

[1987_6] F. J. Pedrotti and L. S. Pedrotti, *Introduction to optics*, Prentice-Hall, 1987, pp. 391-404.

[1987_7] A. Sugimura, Y. Fukuda, and M. Hanabusa, Selective area deposition of silicon-nitride and silicon-oxide by laser chemical vapor deposition and fabrication of microlenses, J. Appl. Phys. **62**, 3222-3227 (1987).

[1987_8] J. Durnin, Exact solution for nondiffracting beams I - The scalar theory, J. Opt. Soc. Am. A **4**, 651-654 (1987).

[1987_9] J. Durnin, J. J. Miceli, and J. H. Eberly, Diffraction-free beams, Phys. Rev. Lett. **58**, 1499-1501 (1987).

[1988_1] J. Durnin, J. J. Miceli Jr., and J. H. Eberly, Comparison of Bessel and Gaussian beams, Opt. Lett. **13**, 79-80 (1988).

[1988_2] J. Turunen, A. Vasara, and A. T. Friberg, Holographic generation of diffraction-free beams, Appl. Opt. **27**, 3959-3962 (1988).

[1988_3] M. Oikawa, E. Okuda, K. Hamanaka, and H. Nemoto, Integrated planar microlens and its applications, Proc. SPIE Vol. **898**, 3-10 (1988).

[1988_4] Z. D. Popovic, R. A. Sprague, and G. A. N. Connell, Technique for monolithic fabrication of microlens arrays, Appl. Opt. **27**, 1281-1284

(1988).

[1988_5]   A. W. Lohmann, An array illuminator based on the Talbot-effect, Optik **79**, 41-45 (1988).

[1988_6]   J. R. Leger, M. L. Scott, and W. B. Veldkamp, Coherent operation of AlGaAs lasers using microlenses and diffractive coupling, Appl. Phys. Lett. **52**, 1771-1773 (1988).

[1988_7]   D. Mehuys, K. Mitsunaga, L. Eng, W. K. Marshall, and A. Yariv, Supermode control in diffraction-coupled semiconductor laser arrays, Appl. Phys. Lett. **53**, 1165-1167 (1988).

[1988_8]   M. Land, The optics of animal eyes, Contemp. Phys. **29**, 435-455 (1988).

[1988_9]   C. Zizzo, C. Arnone, C. Cali, and S. Sciortino, Fabrication and characterization of Gaussian mirrors for the visible and the near Infrared, Opt. Lett. **13**, 342-344 (1988).

[1988_10] K. J. Snell, N. McCarthy, M. Piché, and P. Lavigne, Single transverse mode oscillation from an unstable resonator Nd:YAG-laser using a variable reflectivity mirror, Opt. Commun. **65**, 377-382 (1988).

[1988_11] S. DeSilvestri, P. Laporta, V. Magni, and O. Svelto, Solid-state laser unstable resonator with tapered reflectivity mirrors: the super-Gaussian apprach, IEEE J. Quant. Electron. **24**, 1172-1177 (1988).

[1988_12] S. DeSilvestri, P. Laporta, V. Magni, O. Svelto, C. Arnone, C. Cali, F. Sciortino, and C. Zizzo, Nd:YAG-laser with multidielectric variable reflectivity output coupler, Opt. Commun. **67**, 229-232 (1988).

[1988_13] D. V. Willetts and M. R. Harris, Output characteristics of a compact 1 J carbon dioxide laser with a Gaussian reflectivity resonator, IEEE J. Quant. Electron. **24**, 849-855 (1988).

[1988_14] A. Thelen, *Design of optical interference coatings*, McGraw-Hill, New York, 1988.

[1989_1]   A. W. Lohmann, Scaling laws for lenses, Appl. Opt. **28**, 4996-4998 (1989).

[1989_2]   J. Bennett and L. Mattson, *Introduction to surface roughness and scattering*, Optical Society of America, Washington, D.C., 1989.

[1989_3]   M. Born and E. Wolf, *Principles of optics - electromagnetic theory of propagation, interference and diffraction of light*, Pergamon Press, Oxford, 1989.

[1989_4]   R. Grunwald, G. Szczepanski, I. Pinz, and D. Schäfer, Variable-reflectivity IR-laser outcoupling mirrors, Europhysics Conference Abstracts **Vol. 13D** (1989), Part II: Posters P 2.31.

[1989_5]   G. Indebetouw, Nondiffracting optical fields: some remarks on their analysis and synthesis, J. Opt. Soc. Am. A **6**, 150-157 (1989).

[1989_6]   R. W. Ziolkowski, Localized transmission of electromagnetic energy, Phys. Rev. A **39**, 2005-2033 (1989).

[1989_7] L. M. Soroko, Axicons and meso-optical imaging devices, in: E. Wolf (Ed.), *Progress in Optics*, North-Holland, Amsterdam, 1989, pp. 109-160.

[1989_8] V. M. Weerasinghe, J. Gabzdyl, and R. D. Hibberd, Properties of a laser beam generated from an axicon resonator, Opt. Laser Technol. **21**, 339-342 (1989).

[1989_9] S. R. Jahan and M. A. Karim, Refracting systems for Gaussian-to-uniform beam transformations, Optics & Laser Technology **21**, 27-30 (1989).

[1989_10] J. M. Finlan and K. M. Flood, Collimation of diode laser arrays using etched cylindrical computer-generated holographic lenses, Proc. SPIE Vol. **1052**, 186-190 (1989).

[1989_11] S. Ogata, H. Sekii, T. Maeda, H. Goto, T. Yamashita, and K. Imanaka, Microcollimated laser diode with low wavefront aberration, IEEE Photonics Technology Letters **1**, 354-355 (1989).

[1989_12] M. Tanigami, S. Ogata, S. Aoyama, T. Yamashita, and K. Imanaka, Low-wavefront aberration and high-temperature stability molded micro Fresnel lens, IEEE Photonics Technology Letters **1**, 384-385 (1989).

[1989_13] K. Patorski, The self-imaging phenomenon and its applications, in E. Wolf (Ed.), *Progress in Optics*, Vol. **27**, North-Holland, Amsterdam, 1989, pp. 101-110.

[1989_14] F. X. D'Amato, E. T. Siebert, and C. Roychoudhuri, Coherent operation of diode lasers using a spatial filter in a Talbot cavity, Appl. Phys. Lett. **55**, 816-818 (1989).

[1989_15] W. Goltos and M. Holz: Binary micro-optics, an application to beam steering, SPIE Vol. **1052**, 131 (1989).

[1989_16] Y. Ozaki and K. Takamoto, Cylindrical fly's eye lens for intensity redistribution of an excimer laser beam, Appl. Opt. **28**, 106-110 (1989).

[1989_17] C. Grèzes-Besset, R. Richier, and E. Pelletier, Layer uniformity obtained by vacuum evaporation. Application to Fabry-Perot filters, Appl. Opt. **28**, 2960-2964 (1989).

[1989_19] A. Piegari, A. Tirabassi, and G. Emiliani, Thin films for special laser mirrors with radially variable reflectance: production techniques and laser testing, Proc. SPIE Vol. **1125**, paper 1125-12 (1989).

[1989_20] S. De Silvestri, V. Magni, and O. Svelto, Modes of resonators with mirror reflectivity modulated by absorbing masks, Appl. Opt. **28**, 3684-3690 (1989).

[1989_21] J. R. Leger, Lateral mode control of an AlGaAs laser array in a Talbot cavity, Appl. Phys. Lett. **55**, 334-336 (1989).

[1990_1] J. Jahns and S. Walker, Two-dimensional array of diffractive microlenses fabricated by thin-film deposition, Appl. Opt. **29**, 931-936 (1990).

[1990_2] M. Kubo and M. Hanabusa, Fabrication of microlenses by laser chemical vapor deposition, Appl. Opt. **29**, 2755-2759 (1990).

[1990_3] M. Agu, A. Akiba, T. Mochizuki, and S. Kamemaru, Multimatched filtering using a microlens array for an optical-neural pattern recognition system, Appl. Opt. **29**, 4087-4091 (1990).

[1990_4] X. Yang, T. Lu, and F. T. S. Yu, Compact optical neural network using cascaded liquid crystal televisions, Appl. Opt. **29**, 5223-5225 (1990).

[1990_5] J. Jahns and W. Däschner, Optical cyclic shifter using diffractive lenslet arrays, Opt. Commun. **79**, 407-410 (1990).

[1990_6] D. J. Shaw and T. A. King, Densification of sol-gel silica glass by laser irradiation, Proc. SPIE Vol. **1328**, 474-481 (1990).

[1990_7] V. P. Veiko, E. B. Yakovlev, G. K. Kostyuk, P. A. Fomichev, V. A. Chuiko, and V. S. Kozhukharov, New technology of optical components based on local laser thermo-consolidation of porous glasses and coats, Proc. SPIE Vol. **1328**, 201-205 (1990).

[1990_8] J. A. Dobrowolski and R. A. Kemp, Refinement of optical multilayer systems with different optimisation procedures, Appl. Opt. **29**, 2876-2893 (1990).

[1990_9] R. Grunwald, I. Pinz, H. Schönnagel, and D. Schäfer, Apodizing IR-laser outcoupling mirrors: design, fabrication and application, Akademie der Wissenschaften der DDR, Zentralinstitut für Optik und Spektroskopie, Preprint 90-4, 1990.

[1990_10] S. G. Lukishova and S. A. Kovtonuk, Dielectric films deposition with cross-section variable thickness for amplitude filters on the basis of frustrated total internal reflection, Proc. SPIE Vol. **1270**, 260-271 (1990).

[1990_11] E. Ho, F. Koyama, and K. Iga, Effective reflectivity from self-imaging in a Talbot cavity and its effect on the threshold of a finite 2-D surface emitting laser array, Appl. Opt. **29**, 5080-5085 (1990).

[1990_12] Yu. T. Mazurenko, Holography of wave packets, Appl. Phys. B **50**, 101-113 (1990).

[1991_1] H.-D. Wu and F. S. Barnes (Eds.), *Microlenses - coupling light to optical fibers*, Progress in Lasers and Electro-optics, IEEE Press, New York 1991.

[1991_2] M. C. Hutley, D. Daly, and R. F. Stevens, The testing of microlens arrays, IOP Short Meeting Series, Vol. **30**, pp. 67-81 (1991).

[1991_3] Ch. Budzinski, R.Grunwald, I. Pinz, D. Schäfer, and H. Schönnagel, Apodized outcouplers for unstable resonators, Proc. SPIE Vol. **1500**, 264-274 (1991).

[1991_4] R. Grunwald, H. Schönnagel, and I. Pinz, Lasers with apodizing dielectric components: influence of the mirror orientation on the beam properties, Laser 1991 LASERION, June 12-14, München 1991, Book of Abstracts, 217.

[1991_5] R. M. Herman and T. A. Wiggins, Production and uses of diffractionless beams, J. Opt. Soc. Am. A **8**, 932-942 (1991).

[1991_6] F. Bloisi and L. Vicari, Bessel beams propagation through axisymmetric optical systems, J. Optics (Paris) **22**, 3-5 (1991).

[1991_7] A. J. Cox and D. C. Dibble, Holographic reproduction of a diffraction-free beam, Appl. Opt. **30**, 1330-1332 (1991).

[1991_8] S. R. Mishra, A vector wave analysis of a Bessel beam, Opt. Commun. **85**, 159-161 (1991).

[1991_10] K. Thewes, M. A. Karim, and A. A. S. Awwal, Diffraction-free beam generation using refractive systems, Optics & Laser Technology **23**, 105-108 (1991).

[1991_11] J. J. Snyder, P. Reichert, and T. M. Baer, Fast diffraction-limited cylindrical microlenses, Appl. Opt. **30**, 2743-2747 (1991).

[1991_12] M. Hutley, R. Stevens, and D. Daly, Microlens arrays, Physics World, July 1991, 27-32 (1991).

[1991_13] D. N. Qu, R. E. Burge, and X. Yuan, Diffractive properties of surface relief micro-structures, Proc. SPIE Vol. **1506**, 152-159 (1991).

[1991_14] J. Göttert and J. Mohr, Characterization of micro-optical components fabricated by deep-etch X-ray lithography, Proc. SPIE Vol. **1506**, 170-178 (1991).

[1991_15] N. F. Borelli, R. H. Bellman, J. A. Durbin, and W. Lama, Imaging and radiometric properties of microlens arrays, Appl. Opt. **30**, 3633-3642 (1991).

[1991_16] M. Zoboli and P. Bassi, Experimental characterization of microlenses for WDM transmission systems, Proc. SPIE Vol. **1506**, 160-169 (1991).

[1991_17] M. T. Gale, G. K. Lang, J. M. Raynor, and H. Schuetz, Fabrication of micro-optical components by laser beam writing in photoresist, Proc. SPIE Vol. **1506**, 65-70 (1991).

[1991_18] N. J. Phillips and Ch. A. Barnett, Micro-optic studies using photopolymers, Proc. SPIE Vol. **1544**, 10-21 (1991).

[1991_20] K.-H. Brenner, 3D-Integration of Optical Systems, Proc. SPIE Vol. **1506**, 94-98 (1991).

[1991_21] J. Bartley and W. Goltoos, Laser ablation of refractive micro-optic lenslet arrays, Proc. SPIE Vol. **1544**, 140-145 (1991).

[1991_22] V. P. Veiko, E. B. Yakovlev, V. V. Frolov, V. A. Chuiko, A. K. Kromin, M. O. Abbakumov, A. T. Shakola, and P. A. Fomichev, Laser heating and evaporation of glass and glass-borning materials and its application for creating MOC, Proc. SPIE Vol. **1544**, 152-163 (1991).

[1991_23] F. T. S. Yu, X. Yang, S. Yin, and D. A. Gregory, Mirror-array optical interconnected neural network, Opt. Lett. **16**, 1602-1604 (1991).

[1991_24] R. Grunwald, H. Schönnagel, B. Voigt, J. Guhr, A. Mitreiter, I. Pinz, and J. Bleck, Optimization of a coherent TE-CO2-laser with variable reflectivity outcouplers, Infrared Phys. **32**, 109-127 (1991).

[1991_25] N. Streibl, U. Nölscher, J. Jahns, and S. J. Walker, Array generation with lenslet arrays, Appl. Opt. **30**, 2739-2742 (1991).

[1991_26] M. O. Sculli and M. S. Zubairy, Simple laser accelerator: optics and particle dynamics, Phys. Rev. **A 44**, 2656-2663 (1991).

[1992_1] W. B. Veldkamp and T. J. McHugh, Binary optics, Scientific American **266**, 92-97 (1992).

[1992_2] J. Y. Lu and J. F. Greenleaf, Nondiffracting X-waves. Exact solutions to free space scalar wave equation and their finite aperture realizations, IEEE Trans. Ultrason. Ferroelec. Freq. Cont. **39**, 19-31 (1992).

[1992_3] J. Y. Lu and J. F. Greenleaf, Experimental verification of nondiffracting X waves, IEEE Trans. Ultrason. Ferroelec. Freq. Cont. **39**, 441-446 (1992).

[1992_4] J. Jahanmir and J. C. Wyant, Comparison of surface roughness measured with an optical profiler and a scanning probe microscope, Proc. SPIE Vol. **1720**, 111- 118 (1992).

[1992_5] A. Akhmanov, V. A. Vysloukh, and A. S. Chirkin, *Optics of Femtosecond Laser Pulses*, American Institute of Physics, New York, 1992, pp. 51-53.

[1992_6] D. Malacara (Ed.), *Optical shop testing*, John Wiley & Sons, New York 1992.

[1992_7] M. M. Wefers and K. A. Nelson, Programmable phase and amplitude femtosecond pulse shaping, Opt. Lett. **18**, 2032-2034 (1992).

[1992_8] N. C. Roberts, A. G. Kirk, and T. J. Hall, Binary phase gratings for hexagonal array generation, Opt. Comm. **94**, 501-505 (1992).

[1992_9] F. B. McCormick, F. A. P. Tooley, T. J. Cloonan, J. M. Sasian, H. S. Hinton, K. O. Mersereau, and A. Y. Feldblum, Optical interconnections using microlens arrays, Opt. Quant. Electron. **24**, S465-S477 (1992).

[1992_10] P. Latimer and R. F. Course, Talbot effect reinterpreted, Appl. Opt. **31**, 80-89 (1992).

[1992_11] G. Artzner, Microlens arrays for Shack-Hartmann wavefront sensors, Opt. Eng. **31**, 1311-1322 (1992).

[1992_12] M. T. Gale, C. Appassito, G. K. Lang, J. S. Pederson, J. M. Raynor, H. Schütz, D. Prongué, and P. Ehbets, Microoptical components, Paul Scherrer Institut (PSI) Zürich, Jahresbericht 1992, p. 28.

[1992_13] G. Scott and K. Henry, Excimer laser processing of aerospace alloys, Proc. SPIE Vol. **1835**, 199-126 (1992).

[1992_14] B. D. Fabes, B. J. J. Zelinski, D. J. Taylor, and L. Weisenbach, Laser densification of optical films, Proc. SPIE Vol. **1758**, 227-234 (1992).

[1992_15] T. Chia, J. K. West, and L. L. Hench, Fabrication of micro lenses by laser densification on gel silica glass, in: L. L. Hench and J. K. West (Eds.), *Chemical Processing of Advanced Materials*, John Wiley&Sons, New York, 1992, pp. 933-939.

[1992_16] H. Suda and M. Hanabusa, Fabrication of microlenses by laser-induced vaporization, Appl. Opt. **31**, 5388-5390 (1992).

[1992_17] K. M. Abramski, H. J. Baker, and D. R. Hall, Single-mode selection using coherent imaging within a slab waveguide CO2 laser, Appl. Phys. Lett. **60**, 2469-2471 (1992).

[1993_1] W. Ehrfeld, G. Wegner, W. Karthe, H.-D. Bauer, and H. O. Moser (Eds.), *Integrated optics and micro-optics with polymers*, B. G. Teubner Verlagsgesellschaft, Stuttgart, Leipzig, 1993.

[1993_2] E. Gratix, Evolution of a microlens surface under etching conditions, Proc. SPIE Vol. **1992**, paper 266-74 (1993).

[1993_3] S. Haselbeck, H. Schreiber, J. Schwider, and N. Streibl, Microlenses fabricated by melting photoresist, Opt. Eng. **6**, 1322-1324 (1993).

[1993_4] R. Grunwald, U. Griebner, and D. Schäfer, Graded reflectivity micro-mirror arrays, Proc. SPIE Vol. **1983** (1993), 49-50.

[1993_5] R. Grunwald and U. Griebner, Optically segmented unstable resonators with graded reflectance micro-mirror arrays, Workshop laser resonators with graded reflectance mirrors, Florence, Italy, 8-9 September, 1993, Technical Abstracts.

[1993_6] J. M. Bennett and L. Mattson, *Surface roughness and scattering*, Optical Society of America, Washington, D.C., 1993.

[1993_7] M. E. Motamedi, R. J. Anderson, R. de la Rosa, L. G. Hale, W. J. Gunning III, R. L. Hall, and M. Khoshnevisan, Binary optics thin-film microlens array, Proc. SPIE Vol. **1751**, 22-23 (1993).

[1993_8] K. M. Iftekharuddin, A. A. S. Awwal, and M. A. Karim, Gaussian-to-Bessel beam transformation using a split refracting system, Appl. Opt. **32**, 2252-2256 (1993).

[1993_9] Z. Bouchal, Dependence of Bessel beam characteristics on angular spectrum phase variations, J. Mod. Opt. **40**, 1325-1329 (1993).

[1993_10] J. J. Snyder and A. E. Cable, Cylindrical microlenses improve laser diode beams, Laser Focus World, 97-100 (Feb. 1993).

[1993_11] D. Mendlovic, Three-dimensional image sensing based on a zone-plate array, Opt. Commun. **95**, 26-32 (1993).

[1993_12] S. Mihailov and S. Lazare, Fabrication of refractive microlens arrays by excimer laser ablation of amorphous Teflon, Appl. Opt. **32**, 6211-6218 (1993).

[1993_13] K.-H. Brenner, M. Kufner, S. Kufner, J. Moisel, A. Müller, S. Sinzinger, M. Testorf, J. Göttert, and J. Mohr, Application of three-dimensional microoptical components formed by lithography, electroforming, and plastic molding, Appl. Opt. **32**, 6464-6469 (1993).

[1993_14] E. Noponen, J. Turunen, and A. Vasara, Electromagnetic theory and design of diffractive-lens arrays, J. Opt. Soc. Am. A **10**, 434-443 (1993).

[1993_15] V. Arrizón and J. Ojeda-Castaneda, Talbot array illuminators with binary phase gratings, Opt. Lett. **18**, 1-3 (1993).

[1993_16] P. Szwaykowski, Talbot effect reinterpreted: comment, Appl. Opt. **32**, 3466-3467 (1993).

[1993_17] P. Latimer, Talbot effect reinterpreted: reply to comment, Appl. Opt. **32**, 3468-3469 (1993).

[1993_18] P. Latimer, Use of the Talbot effect to couple the phases of lasers, Appl. Phys. Lett. **62**, 217-218 (1993).

[1993_19] Eyes for Robots, Laser und Optoelektronik **25**, 18 (1993).

[1993_20] M. W. Farn, M. B. Stern, W. B. Veldkamp, and S. S. Medeiros, Color separation by use of binary optics, Opt. Lett. **18**, 1214-1216 (1993).

[1993_21] W. B. Veldkamp, Wireless focal planes, on the road to amacronic sensors, IEEE J. QE **29**, 801-813 (1993).

[1993_22] J. Druessel, J. Grantham, and P. Haaland, Optimal phase modulation for gradient-index optical filters, Opt. Lett. **18**, 1583-1585 (1993).

[1993_23] C. Grèzes-Besset, F. Chazallet, G. Albrand, and E. Pelletier, Synthesis and research of the optimum conditions for the optical monitoring of non-quarter-wave multilayers, Appl. Opt. **32**, 5612-5618 (1993).

[1993_24] A. Yu Okulov, Scaling of diode-array-pumped solid-state lasers via self-imaging, Opt. Commun., **99**, 350-354 (1993).

[1993_25] T. Wulle and S. Hemminghaus, Nonlinear optics of Bessel beams, Phys. Rev. Lett. **70**, 1401-1404 (1993).

[1993_26] K. G. Purchase, D. J. Brady, and K. Wagner, Time-of-flight cross correlation on a detector array for ultrafast packet detection, Opt. Lett. **18**, 2129-2131 (1993).

[1993_27] D. J. Kane and R. Trebino, Single-shot measurement of the intensity and phase of an arbitrary ultrashort pulse by using frequency-resolved optical gating, Opt. Lett. **18**, 823-825 (1993).

[1994_1] S. H. Lee (Ed.), *Diffractive and Miniaturized Optics*, Critical reviews of optical science and technology, Vol. **CR49**, SPIE Optical Engineering Press, Bellingham, Washington, 1994.

[1994_2] R. Grunwald and U. Griebner, Segmented solid-state laser resonators with graded reflectance micro-mirror arrays, Pure Appl. Opt. **3** (1994), 435-440.

[1994_3] R. Grunwald, U. Griebner, and R. Koch, Phase-coupled multiple-beam solid-state laser with Talbot-resonator, CLEO 1994, Anaheim, USA, May 8-13, 1994, paper CFE2, Technical Digest, 410.

[1994_4] S. Ruschin, Modified Bessel nondiffracting beams, J. Opt. Soc. Am. A **11**, 3224-3228 (1994).

[1994_5] J. A. Davis, D. M. Cottrell, C. A. Maley, and M. R. Crivello, Subdiffraction-limited focusing lens, Appl. Opt. **33**, 4128-4131 (1994).

[1994_6] R. Piestun and J. Shamir, Control of wave-front propagation with diffractive elements, Opt. Lett. **19**, 771-773 (1994).

[1994_7] J. Hutfless, T. Rebhahn, N. Lutz, M. Geiger, M. Frank, N. Streibl, and J. Schwider, Mikrooptiken für die effiziente Materialbearbeitung mit Excimerlasern, Laser und Optoelektronik **26**, 50-57 (1994).

[1994_8] M. E. Motamedi, A. P. Andrews, W. J. Gunning, and M. Khoshnevisan, Miniaturized micro-optical scanners, Opt. Eng. **33**, 3616-3623 (1994).

[1994_9] M. E. Motamedi, Micro-opto-electro-mechanical systems, Opt. Eng. **33**, 3505-3517 (1994).

[1994_10] Z. L. Liau, J. N. Waloplwe, D. E. Mull, C. L. Dennis, and L. J. Misaggia, Accurate fabrication of anamorphic microlenses and efficient collimation of tapered unstable-resonator diode lasers, Appl. Phys. Lett. **64**, 3368-3370 (1994).

[1994_11] W. Chen, C. S. Roychoudhuri, and C. M. Banas, Design approaches for laser-diode material-processing systems using fibers and micro-optics, Opt. Eng. **33**, 3662-3669 (1994).

[1994_12] S. Sanders, R. Waarts, D. Nam, D. Welch, D. Scifres, J. C. Ehlert, W. J. Cassarly, J. M. Finlan, and K. M. Flood, High power coherent two-dimensional semiconductor laser array, Appl. Phys. Lett. **64**, 1478-1480 (1994).

[1994_13] H. J. Tiziani and H.-M. Uhde, Three-dimensional analysis by a microlens-array confocal arrangement, Appl. Opt. **33**, 567-572 (1994).

[1994_14] J. Jahns, F. Sauer, B. Tell, K. F. Brown-Goebeler, A. Y. Feldblum, C. R. Nijander, and W. P. Townsend, Parallel optical interconnections using surface-emitting microlasers and a hybrid imaging system, Opt. Commun. **109**, 328-337 (1994).

[1994_15] E. Carcolé, J. Campos, and S. Bosch, Diffraction theory of Fresnel lenses encoded in low-resolution devices, Appl. Opt. **33**, 162-174 (1994).

[1994_16] Z. L. Liau, D. E. Mull, C. L. Dennis, R. C. Williamson, and R. G. Waarts, Large-numerical-aperture microlens fabrication by one-step etching and mass-transport smoothing, Appl. Phys. Lett. **64**, 1484-1486 (1994).

[1994_17] T. R. Jay and M. B. Stern, Preshaping photoresist for refractive microlens fabrication, Opt. Eng. **33**, 3552-3555 (1994).

[1994_18] M. B. Stern and T. R. Jay, Dry etching for coherent refractive microlens arrays, Opt. Eng. **33**, 3547-3551 (1994).

[1994_19] H. M. Ozaktas, H. Urey, and A. W. Lohmann, Scaling of diffractive and refractive lenses for optical computing and interconnections, Appl. Opt. **33**,

3782-3789 (1994).

[1994_20] M. Ferstl, B. Kuhlow, and E. Pawlowski, Effect of fabrication errors on multilevel Fresnel zone lenses, Opt. Eng. **33**, 1229-1235 (1994).

[1994_21] H. Zarschizky, A. Stemmer, F. Mayerhofer, G. Lefranc, and W. Gramann, Binary and multilevel diffractive lenses with submicrometer feature sizes, Opt. Eng. **33**, 3527-3536 (1994).

[1994_22] R. Wälti, W. Lüthy, and H. P. Weber, Collimation of diode laser arrays with cylinder lenses, Pure Appl. Opt. **3**, 199-208 (1994).

[1994_23] M. H. Maleki and A. J. Devaney, Noniterative reconstruction of complex-valued objects from two intensity measurements, Opt. Eng. **33**, 3243-3253 (1994).

[1994_24] G. Przyrembel, Continous-relief microoptical elements fabricated by laser beam writing, in: H. Reichl and A. Heuberger (Eds.), *Micro System Technologies '94*, VDE-Verlag, Berlin, Offenbach (1994), pp. 219-228.

[1994_25] D L. Kendall, W. P. Eaton, R. Manginell, and T. G. Digges, Jr., Micromirror arrays using KOH:H$_2$O micromachining of silicon for lens templates, geodesic lenses, and other applications, Opt. Eng. **33**, 3578-3588 (1994).

[1994_26] M. F. Lewis and R. A. Wilson, The use of lenslet arrays in spatial light modulators, Pure Appl. Opt. **3**, 143-150 (1994).

[1994_27] R. Szipöc, K. Ferencz, C. Spielmann, and F. Krausz, Chirped multilayer coatings for broad-band dispersion control in femtosecond lasers, Opt. Lett. **19**, 201-203 (1994).

[1994_28] E. Bonet, P. Andrés, J. C. Barreiro, and A. Pons, Self-imaging properties of a periodic microlens array: versatile array illuminator realization, Opt. Commun. **106**, 39-44 (1994).

[1994_29] K. Zimmer and F. Bigl: 3D-Strukturierung von Polymeren durch Excimerlaserablation, 1. Int. Mittweidaer Fachtagung Qualitäts- und Informationsmanagement, 1994.

[1994_30] Microlens Arrays, Topical Meeting, 13-14 May 1993, National Physical Laboratory, Teddington, UK, in: Pure Appl. Opt. **3** (1994).

[1994_31] S. Ogata, J. Ishida, and T. Sasano, Optical sensor array in an artificial compound eye, Opt.Eng. **33**, 3649-3655 (1994).

[1994_32] J. Shimada, O. Ohguchi, and R. Sawada: Focusing characteristics of a wide-striped laser diode integrated with a microlens, J. Lightwave Technol. **12**, 936-942 (1994).

[1994_33] E. Pawlowski, H. Engel, M. Ferstl, W. Fürst, and B. Kuhlow, Diffractive microlenses with antireflection coatings fabricated by thin-film deposition, Opt. Eng. **33**, 647-652 (1994).

[1994_34] E. Pawlowski and B. Kuhlow, Antireflection-coated diffractive optical elements fabricated by thin-film deposition, Opt. Eng. **33**, 3537-3546 (1994).

[1994_35] E. Carcolé, J. Campos, and S. Bosch, Diffraction Theory of Fresnel lenses encoded in low-resolution devices, Appl. Opt. **33** (1994), 162-174.

[1994_36] D. L. MacFarlane, V. Narayan, J. A. Tatum, W. R. Cox, T. Chen, and D. J. Hayes, Microjet fabrication of microlens arrays, IEEE Photonics Technol. Lett. **6**, 1112-1114 (1994).

[1994_37] V. P. Veiko and Y. B. Yakovlev, Physical fundamentals of laser forming of micro optical components, Opt. Eng. **33**, 3567-3571 (1994).

[1994_38] A. Premoli and M. L. Rastello, Minimax refining of wideband antireflection coatings for wide angular incidence, Appl. Opt. **33**, 2018-2024 (1994).

[1994_39] H. Sickinger, O. R. Falkenstoerfer, N. Lindlein, and J. Schwider, Characterization of microlenses using a phase-shifting shearing interferometer, Opt. Eng. **33**, 2680-2686 (1994).

[1994_40] I. A. Walmsley and V. Wong, Analysis of ultrashort pulse-shape measurement using linear interferometers, Opt. Lett. **19**, 287–289 (1994).

[1995_1] S. Sinzinger, K.-H. Brenner, J. Moisel, T. Spick, and M. Testorf, Astigmatic gradient-index elements for laser-diode collimation and beam shaping, Appl. Opt. **29**, 6626-6632 (1995).

[1995_2] R. Grunwald, U. Griebner, R. Ehlert, and S. Woggon, Micro-optical array components for novel-type lasers and artificial compound eyes, National Workshop on Microlens Arrays, National Physical Laboratory, Teddington, UK, 11-12 May 1995, Paper A4, pp. 85-88.

[1995_3] R. Grunwald and U. Griebner, Passively Q-switched high-power Nd:glass laser with different types of graded reflectance outcoupling mirrors, CLEO 1995, Baltimore 1995, paper CMI3, Technical Digest, 38-39.

[1995_4] A. Zhiping, Q. Lu, and Z. Liu, Propagation of apertured Bessel beams, Appl. Opt. **34**, 7183-7185 (1995).

[1995_5] W. Singer, M. Testorf, and K.-H. Brenner, Gradient-index microlenses: numerical investigation of different spherical index profiles with the wave propagation method, Appl. Opt. **34**, 2165-2171 (1995).

[1995_6] A. Rohrbach and K.-H. Brenner, Surface-relief phase structures generated by light-initiated polymerization, Appl. Opt. **34**, 4747-4754 (1995).

[1995_7] M. Testorf and S. Sinzinger, Evaluation of microlens properties in the presence of high spherical aberration, Appl. Opt. **34**, 6431-6437 (1995).

[1995_8] V. Kabelka and A. V. Masalov, Time-frequency imaging of a single ultrashort light pulse from angularly resolved autocorrelation, Opt. Lett. **20**, 1301-1303 (1995).

[1995_9] M. Rossi, R. E. Kunz, and H. P. Herzig, Refractive and diffractive properties of planar micro-optical elements, Appl. Opt. **34**, 5996-6007 (1995).

[1995_10] M. Kuittinen, H. P. Herzig, and P. Ehbets, Improvements in diffraction efficiency of gratings and microlenses with continuous relief structures, Opt. Commun. **120**, 230-234 (1995).

[1995_11] X. Tan, B.-Y. Gu, G.-Z. Yang, and B.-Z. Dong, Diffractive phase elements for beam shaping: a new design method, Applied Optics **34**, 1314-1320 (1995).

[1995_12] M. Kufner, S. Kufner, P. Chavel, and M. Frank, Monolithic integration of microlens arrays and fiber holder arrays in poly(methyl methacrylate) with fiber self-centering, Opt. Lett. **20**, 276-278 (1995).

[1995_13] S. H. Song and E. H. Lee, Planar optical configurations for crossover interconnects, Opt. Lett. **20**, 617-619 (1995).

[1995_14] S. Sinzinger and M. Testorf, Transition between diffractive and refractive micro-optical components, Appl. Opt. **34**, 5970-5976 (1995).

[1995_15] T. S. Rose, D. A. Hinkley, and R. A. Fields, Efficient collection and manipulation of laser diode output using refractive micro-optics, Proc. SPIE Vol. **2383**, 273-377 (1995).

[1995_16] J. M. Geary, *Introduction to wavefront sensors*, SPIE Optical Engineering Press, Bellingham, 1995.

[1995_17] B. M. Welsh, B. L. Ellenbroek, M. C. Roggemann, and T. L. Pennington, Fundamental performance comparison of a Hartmann and a shearing interferometer wavefront sensor, Appl. Opt. **34**, 4186-4195 (1995).

[1995_18] M. E. Motamedi and L. Beiser (Eds.), Micro-optics, micromechanics and laser scanning and shaping, Proc. SPIE Vol. **2383** (1995).

[1995_19] K. Engelhardt and K. Knop, Passive focus sensor, Appl. Opt. **34**, 2339-2344 (1995).

[1995_20] J.-K. Ji and Y.-S. Kwon, Conical microlens arrays that flatten optical-irradiance profiles of nonuniform sources, Appl. Opt. **34**, 2841-2843 (1995).

[1995_21] T. H. Bett, C. N. Danson, P. Jinks, D. A. Pepler, I. N. Ross, and R. M. Stevenson, Binary phase zone-plate arrays for laser-beam spatial-intensity distribution conversion, Appl. Opt. **34**, 4025-4036 (1995).

[1995_22] I. Matsushima, T. Tomie, Y. Matsumoto, I. Okuda, E. Miura, H. Yashiro, Eiichi Takahashi, and Y. Owadano, Two-dimensional beam smoothing by broadband random-phase irradiation, Opt. Comm. **120**, 299-302 (1995).

[1995_23] W. Bacher, W. Menz, and J. Mohr, The LIGA technique and its potential for microsystems - a survey, IEEE Transactions on Industrial Electronics **42**, 431-441 (1995).

[1995_24] H. Hisakuni and K. Tanaka, Optical fabrication of microlenses in chalcogenide glasses, Opt. Lett. **20** , 958-960 (1995).

[1995_25] J. Ihlemann and B. Wolff-Rottke, Excimer laser patterning of dielectric layers, Appl. Surface Sci. **86**, 228-233 (1995), 228-233.

[1995_26] F. R. Flory (Ed.), *Thin films for optical systems*, Marcel Dekker, New York, 1995.

[1995_27] V. N. Mahajan, Zernike polynomials and optical aberrations, Appl. Opt. **34**,

8060-8062 (1995).

[1995_28] P. G. Verly, Fourier transform technique with frequency filtering for optical thin-film design, Appl. Opt. **34**, 688-694 (1995).

[1995_29] C. J. van der Laan and H. J. Frankena, Equivalent layers: another way to look at them, Appl. Opt. **34**, 681-687 (1995).

[1995_30] H. Stöcker (Ed.), *Taschenbuch mathematischer Formeln und moderner Verfahren*, Verlag Harry Deutsch, Frankfurt am Main, 1995 (in German).

[1995_31] R. Grunwald, U. Griebner, and R. Ehlert, Microlens arrays for segmented laser architectures, Proc. SPIE Vol. **2383**, 324-333 (1995).

[1995_32] U. Griebner, H. Schönnagel, and R. Grunwald, Diode-pumping of fiber array lasers via microlens arrays, in: B. H. T. Chai, S. A. Payne (Eds.), OSA Proceedings on Advanced Solid-State Lasers, **Vol. 24**, 253-256 (1995).

[1995_33] J. Schwider and O. Falkenstoerfer, Twyman-Green interferometer for testing microspheres, Opt. Eng. **34**, 2972-2975 (1995).

[1995_34] W. Singer, *Development and experimental verification of non-paraxial wave propagation methods for the analysis of microoptical components*, Thesis, Friedrich-Alexander-University, Erlangen-Nürnberg, 1995 (in German).

[1995_35] P. Haberäcker, *Praxis der digitalen Bildverarbeitung und Mustererkennung*, Carl Hanser, München, 1995 (in German).

[1996_1] E. B. Kley, B. Schnabel, H. Hübner, and U. D. Zeitner, Application of metallic subwavelength gratings for polarization devices, Proc. SPIE Vol. **2863**, 166-174 (1996).

[1996_2] R. Grunwald, R. Ehlert, S. Woggon, and H.-H. Witzmann, Thin-film microlens arrays on flexible polymer substrates, in: H. Reichl, A. Heuberger (Eds.), *Micro System Technologies*, VDE-Verlag, Berlin 1996, pp. 793-795.

[1996_3] R. Grunwald, D. Schäfer, R. Ehlert, S. Woggon, and H.-H. Witzmann, Herstellung von Mikrolinsen-Arrays mit hohem Füllfaktor durch gekreuztes Aufdampfen linearer Arrays, 97. Jahrestagung der DGaO, 28.5.-1.6.1996, Neuchâtel, Schweiz, P 13, Abstracts, 102 (in German).

[1996_4] R. Grunwald, R. Ehlert, S. Woggon, H.-J. Pätzold, and H.-H. Witzmann, Microlens arrays formed by crossed thin-film deposition of cylindrical microlenses, Diffractive Optics and Microoptics Topical Meeting, OSA, Boston April 29-May 2, 1996 Technical Digest Series Vol. 5, DMB2, 27-30.

[1996_5] J.-C. Diels and W. Rudolph, *Ultrashort Laser Pulse Phenomena*, Academic Press, San Diego, 1996.

[1996_6] Z. Jaroszewicz, A. Kolodziejczyk, A. Kujawski, and C. Gomez-Reino, Diffractive patterns of small cores generated by interference of Bessel beams, Opt. Lett. **21**, 839-841 (1996).

[1996_7] Z. L. Horváth, J. Vinkó, Zs. Bor, and D. van der Linde, Acceleration of femtosecond pulses to superluminal velocities by Gouy phase shift, Appl. Phys. **B 63**, 481-484 (1996).

[1996_8]   J. Fagerholm, A. T. Friberg, J. Huttunen, D. P. Morgan, and M. M. Salomaa, Angular-spectrum representation of nondiffracting X waves, Phys. Rev. E **54**, 4347-4352 (1996).

[1996_9]   Z. Jaroszewicz, J. F. Román Dopazo, and C. Gomez-Reino, Uniformization of the axial intensity of diffraction axicons by polychromatic illumination, Appl. Opt. **35**, 1025-1031 (1996).

[1996_10] J. Bajer and R. Horák, Nondiffractive fields, Phys. Rev. E **54**, 3052-3054 (1996).

[1996_11] W. Singer, H. P. Herzig, M. Kuittinen, E. Piper, and J. Wangler, Diffractive beamshaping elements at the fabrication limit, Opt. Eng. **35**, 2779-2787 (1996).

[1996_12] H. J. Tiziani, R. Achi, R. N. Krämer, and L. Wiegers, Theoretical analysis of confocal microscopy with microlenses, Appl. Opt. **35**, 120-125 (1996).

[1996_13] S. Lazare, J. Lopez, J.-M. Turlet, M. Kufner, S. Kufner, and P. Chavel, Microlenses fabricated by ultraviolet excimer laser radiation of poly(methyl methacrylate) followed by styrene diffusion, Appl. Opt. **35**, 4471-4475 (1996).

[1996_14] S. Calixto and M. S. Scholl, Relief optical microelements fabricated with dichromated gelatine, Appl. Opt. **36**, 2101-2106 (1996).

[1996_15] S. Calixto and G. P. Padilla, Micromirrors and microlenses fabricated on polymer materials by means of infrared radiation, Appl. Opt. **35**, 6126-6130 (1996).

[1996_16] J. D. Rancourt, Optical thin films - user handbook, SPIE Optical Engineering Press, Bellingham,Washington, 1996.

[1996_17] R. Völkel, H. P. Herzig, Ph. Nussbaum, and R. Dändliker, Microlens array imaging system for photolithography, Opt. Eng. **35**, 3323-3330 (1996).

[1996_18] M. V. Berry and S. Klein, Integer, fractional and fractal Talbot effect, J. Mod. Opt. **43**, 2139-2164 (1996).

[1996_19] I. A. Walmsley and V. Wong, Characterization of the electric field of ultrashort optical pulses, J. Opt. Soc. Am. **B 13**, 2453-2463 (1996).

[1996_20] J. S. Sanders and B. J. Thompson (Eds.), *Selected papers on natural and artificial compound eye sensors*, SPIE Milestone Series Vol. **MS 122**, SPIE Optical Engineering Press, Bellingham, Washington, 1996.

[1996_21] Diffractive Optics and Micro-Optics, OSA Topical Meeting, April 29 - May 2, 1996, Boston, 1996 Technical Digest Series Vol. 5.

[1996_22] D. J. Brink and M. E. Lee, Ellipsometry of diffractive insect reflectors, Appl. Opt. **35**, 1950-1955 (1996).

[1996_23] K. Hamanaka and H. Koshi, An artificial compound eye using a microlens array and its application to scale-invariant processing, Opt. Rev. **3**, 264-268 (1996).

[1996_24] A. Piegari, Coatings with graded-reflectance profile: conventional and unconventional characteristics, Appl. Opt. **35**, 5509-5519 (1996).

[1996_25] L. Li and J. A. Dobrowolski, Visible broadband, wide-angle, thin-film multilayer polarizing beam splitter, Appl. Opt. **35**, 2221-2225 (1996).

[1996_26] Z. Dai, R. Michalzik, P. Unger, and K. J. Ebeling, Impact of the residual facet reflectivity on beam profile filamentation in semiconductor laser amplifiers, CLEO '96, Anaheim 1996, Technical Digest, 416-417.

[1996_27] H. Sõnajalg and P. Saari, Suppression of temporal spread of ultrashort pulses in dispersive media by Bessel beam generators, Opt. Lett. **21**, 1162-1164 (1996).

[1996_28] P. Saari, Spatially and temporally nondiffracting ultrashort pulses, in: O. Svelto, S. De Silvestri, and G. Denardo (Eds.), *Ultrafast Processes in Spectroscopy*, Plenum Press, New York, 1996, 151-156.

[1996_29] P. Hariharan, The Gouy phase shift as a geometrical quantum effect, J. Mod. Opt. **43**, 219-222 (1996).

[1996_30] S. Klewitz, P. Leiderer, S. Herminhaus, and S. Sogomonian, Tunable stimulated Raman scattering by pumping with Bessel beams, Opt. Lett. **21**, 248-250 (1996).

[1996_31] J. A. Davis, E. Carcole, and D. M. Cottrell, Nondiffracting interference patterns generated with programmable spatial light modulator, Appl. Opt. **4**, 599-602 (1996).

[1996_32] J. A. Davis, E. Carcole, and D. M. Cottrell, Intensity and phase measurement of nondiffracting beams generated with a magneto-optic spatial light modulator, Appl. Opt. **35**, 593-598 (1996).

[1996_33] S. De Nicola, On-axis focal shift effects in focused truncated J_0 Bessel beams, Pure Appl. Opt. **5**, 827-831 (1996).

[1996_34] M. M. Wefers and K. A. Nelson, Space-time profiles of shaped ultrafast optical wavefroms, IEEE J. Quant. Electron. **32**, 161-173 (1996).

[1997_1] H.-P. Herzig (Ed.), *Micro-optics*, Taylor & Francis, London 1997.

[1997_2] M. Kufner and S. Kufner, *Micro-optics and lithography*, VUBPRESS - VUB University Press, Brussels, 1997.

[1997_3] M. R. Douglass and Ian S. McMurray, Why is the Texas Instruments Digital Micromirror Device^{TM} (DMD^{TM}) so reliable? DLP White Paper Library, 10/30/1997, Texas Instruments DLP Products, Plano, http://www.dlp.com/dlp_technology/dlp_technology_white_papers.asp.

[1997_4] T. Hessler, M. Rossi, J. Pedersen, T. M. Gale, M. Wegner, D. Steudle, and H. J. Tiziani, Microlens arrays with spatial variation of the optical functions, Pure & Applied Optics **6**, 673-681 (1997).

[1997_5] S. Martellucci and A. N. Chester (Eds.): *Diffractive optics and optical microsystems*, Plenum Press, New York, 1997.

[1997_6] T. Baumert, T. Brixner, V. Seyfried, M. Strehle, and G. Gerber,

Femtosecond pulse shaping by an evolutionary algorithm with feedback, Appl. Phys. B: Lasers and Optics **65**, 779-782 (1997).

[1997_7] D. Mendlovic and A. W. Lohmann, Spacebandwidth product adaptation and its application to superresolution: fundamentals, J. Opt. Soc. Am. **A 14**, 558-562 (1997).
D. Mendlovic and A. W. Lohmann, Spacebandwidth product adaptation and its application to superresolution: experiments, J. Opt. Soc. Am. **A 14**, 563-567 (1997).

[1997_8] R. Grunwald, S. Woggon, R. Ehlert, and W. Reinecke, Thin-film microlens arrays with non-spherical elements, Pure Appl. Opt. **6** (1997), 663-671.

[1997_9] R. Grunwald, S. Woggon, R. Ehlert, and W. Reinecke, Non-spherical microlenses for axial beam shaping, MOC/GRIN 1997, Technical Digest, 372-375.

[1997_10] R. Grunwald, S. Woggon, R. Ehlert, and W. Reinecke, Thin-film microlens arrays with non-spherical elements, Microlens Arrays, EOS Topical Meetings Digest Series **13**, 34-38 (1997).

[1997_11] Th. Hessler, *Continuous-relief diffractive optical elements: design, fabrication, and applications*, PhD Thesis, University of Neuchâtel, 1997.

[1997_12] Z. Jiang, Truncation of a two-dimensional nondiffracting cos beams, J. Opt. Soc. Am. A **14**, 1478-1481 (1997).

[1997_13] R. Borghi and M. Santarsiero, M^2 factor of Bessel-Gauss beams, Opt. Lett. **22**, 262-264 (1997).

[1997_14] P. M. Goorjian and Y. Silberberg, Numerical simulations of light bullets using the full-vector time-dependent nonlinear Maxwell equations, J. Opt. Soc. Am. B **14**, 3253-3260 (1997).

[1997_15] A. T. Friberg, J. Fagerholm, and M. M. Salomaa, Space-frequency analysis of nondiffracting pulses, Opt. Commun. **136**, 207-212 (1997).

[1997_16] L. Erdmann and D. Efferenn, Technique for monolithic fabrication of silicon microlenses with selectable rim angles, Opt. Eng. **36**, 1094-1098 (1997).

[1997_17] A. R. Holdsworth and H. Baker, Assessment of micro-lenses for diode bar collimation, Proc. SPIE Vol. **3000**, 209-214 (1997).

[1997_18] G. M. Peake, S.Z. Sun, and S. D. Hersee, GaAs microlens arrays grown by shadow masked MOVPE, J. Electron. Mat. **26**, 1134-1138 (1997).

[1997_19] P. Heremans, J. Genoe, M. Kuijk, R. Vounckx, and G. Borghs, Mushroom microlenses: optimized microlenses by reflow of multiple layers of photoresist, IEEE Photon. Technol. Lett. **9**, 1367-1369 (1997).

[1997_20] B. P. Keyworth, D. J. Corazza, J. N. McMullin, and L. Mabbott, Single-step fabrication of refractive microlens arrays, Appl. Opt. **36**, 2198-2202 (1997).

[1997_21] S. Glöckner and R. Göring, Investigation of statistical variations between lenslets of microlens arrays, Appl. Opt. **36**, 4438-4445 (1997).

[1997_22] H. Kurita and S. Kawai, Quasi-toric planar microlens for oblique-incidence light beams, Appl. Opt. **36**, 1017-1022 (1997).

[1997_23] P. Coudray, P. Etienne, Y. Moreau, J. Porque, and S. I. Najafi, Sol-gel channel waveguide on silicon: fast direct imprinting and low cost fabrication, Opt. Commun. **143**, 199-202 (1997).

[1997_24] T. R. M. Sales and G. M. Morris, Diffractive refractive behavior of kinoform lenses, Appl. Opt. **36**, 253-257 (1997).

[1997_25] P. C. Sun, Y. T. Mazurenko, and Y. Fainman, Femtosecond pulse imaging: ultrafast optical oscilloscope, J. Opt. Soc. Am. **A 14**, 1159-1170 (1997).

[1997_26] H. Sõnajalg, M. Rätsep, and P. Saari, Demonstration of the Bessel-X pulse propagating with strong lateral and longitudinal localization in a dispersive medium, Opt. Lett. **22**, 310-312 (1997).

[1997_27] F. Nikolajeff, S. Hard, and B. Curtis, Diffractive microlenses replicated in fused silica for excimer laser-beam homogenizing, Appl. Opt. **36**, 8481-8489 (1997).

[1997_28] T. J. Suleski, Generation of Lohmann images from binary-phase Talbot array illuminators, Appl. Opt. **36**, 4686-4691 (1997).

[1997_29] M. F. Land: Microlens arrays in the animal kingdom, Microlens Arrays, EOS Topical Meetings Digest Series **13**, 1-4, NPL Teddington, 15-16 May 1997.

[1997_30] S. H. Song, J.-S. Jeong, S. Park, and E.-H. Lee, Planar optical implementation of fractional correlation, Opt. Commun. **143**, 287-293 (1997).

[1997_31] S. Masuda, S. Takahashi, T. Nose, S. Sato, and H. Ito, Liquid-crystal microlens with a beam-steering function, Appl. Opt. **36**, 4772-4778 (1997).

[1997_32] G. H. Seward and P. F. Mueller, Microlens array structure designed to defeat automated photocopying of documents, Appl. Opt. **22**, , 5363-5371 (1997).

[1997_33] Ch. Hembd, R. Stevens, and M. Hutley, Imaging properties of 'Gabor superlens', Microlens Arrays, European Optical Society Topical Meeting Digest Series **13**, 101-104, NPL Teddington, 15-16 May 1997.

[1997_34] K. Robbie and M. J. Brett, Sculptured thin films and glancing angle deposition: growth mechanisms, J. Vac. Sci. technol. **A 15**, 1460-1465 (1997).

[1997_35] R. Szipöcs and A. Kohazi-Kis, Theory and design of chirped dielectric laser mirrors, Appl. Phys. B **65**, 115-135 (1997).

[1997_36] Y. Zheng, K. Kikuchi, M. Yamasaki, and K. Sonoi, Two-layer wideband antireflection coatings with an absorbing layer, Appl. Opt. **36**, 6335-6338 (1997).

[1997_37] K. V. Popov, J. A. Dobrowolski, A. V. Tikhoravov, and B. T. Sullivan, Broadband high-reflection multilayer coatings at oblique angles of incidence, Appl. Opt. **36**, 2139-2151 (1997).

[1997_38] H. G. Shanbhogue, C. L. Nagendra, M. N. Annapura, S. A. Kumar, and G. K. M. Thutupalli, Multilayer antireflection coatings for the visible and near-infrared regions, Appl. Opt. **25**, 6339-6351 (1997).

[1997_39] M. B. Stern, Pattern transfer for diffractive and refractive microoptics, Microelectron. Eng. **34**, 299-319 (1997).

[1997_40] P. Saari and K. Reivelt, Evidence of X-shaped propagation-invariant localized light waves, Phys. Rev. Lett. **79**, 4135-4138 (1997).

[1997_41] S. Sogomonian, S. Klewitz, and S. Herminghaus, Self-reconstruction of a Bessel beam in nonlinear medium, Opt. Commun. **139**, 313-319 (1997).

[1997_42] W. A. Rodrigues and J. Y. Lu, On the existence of undistorted progressive waves (UPWs) of arbitrary speeds in nature, Found. Phys. **27**, 435-508, 1997.

[1997_43] B. Hafizi, E. Esarey, and P. Sprangle, Laser-driven acceleration with Bessel beams, Phys. Rev. **E 55**, 3539-3545 (1997).

[1997_44] S. Hughes and J. M. Burzler, Theory of Z-scan measurements using Gaussian-Bessel beams, Phys. Rev. **A 56**, R1103-R1106 (1997).

[1997_45] M. K. Pandit and F. P. Payne, Čerenkov second-harmonic generation by nondiffracting Bessel beams in bulk optical crystals, Opt. Quant. Electr. **29**, 35-51 (1997).

[1997_46] M. Erdélyi, Z. L. Horváth, G. Szabó, Zs. Bor, F. K. Tittel, J. R. Cavallaro, and M. C. Smayling, Generation of diffraction-free beams for applications in optical microlithography, J. Vac. Sci. Technol. **B 15**, 287-292 (1997).

[1997_47] E. T. J. Nibbering, O. Dühr, and G. Korn, Generation of intense tunable 20-fs pulses near 400 nm by use of a gas-filled hollow waveguide, Opt. Lett. **22**, 1335-1337 (1997).

[1997_48] C. J. R. Sheppard and X. Gan, Free-space propagation of femto-second light pulses, Opt. Commun **133**, 1-6 (1997).

[1998_1] M. A. Neifield, Information, resolution, and space-bandwidth product, Opt. Lett. **23**, 1477-1479 (1998).

[1998_2] V. N. Mahajan, *Optical imaging and aberrations, Part II: Wave diffraction optics*, SPIE Optical Engineering Press, Bellingham, Washington, 1998.

[1998_3] R. Grunwald, S. Woggon, U. Griebner, R. Ehlert, and W. Reinecke, Axial beam shaping with non-spherical microlenses, Jpn. J. Appl. Phys. **37**, 3701-3707 (1998).

[1998_4] U. Griebner, J. Huschke, R. Grunwald, and H. Schönnagel, Modenselektive Mikrooptiken für miniaturisierte diodengepumpte Festkörperlaser, Verhandlungen der Deutschen Physikalischen Gesellschaft 2/1998, 185 (in German).

[1998_5] H.-J. Pätzold, S. Woggon, R. Grunwald, R. Ehlert, and U. Griebner,

Improvement of a short pulse KrF excimer laser by an unstable resonator with super-Gaussian mirror, CLEO Europe 1998, paper CTuI23, Technical Digest, 81.

[1998_6] C. Rullière (Ed.), *Femtosecond laser pulses*, Springer-Verlag, Berlin, 1998.

[1998_7] S. Chávez-Cerda, E. Tepichin, and M. A. Meneses-Nava, Experimental observation of interfering Bessel beams, Opt. Express **3**, 524-529 (1998).

[1998_8] J. D. Strohschein, H. J. J. Seguin, and C. E. Capjack, Beam propagation constants for a radial laser array, Appl. Opt. **37**, 1045-1048 (1998).

[1998_9] S. Feng, H. G. Winful, and R. W. Hellwarth, Gouy shift and temporal reshaping of focused single-cycle electromagnetic pulses, Opt. Lett. **23**, 385-387 (1998).

[1998_10] R. Piestun and J. Shamir, Generalized propagation-invariant wave fields, J. Opt. Soc. Am. A **15**, 3039-3044 (1998).

[1998_11] Z. Bouchal, J. Wagner, and M. Chlup, Self-reconstruction of a distorted nondiffracting beam, Opt. Commun. **151**, 207-211 (1998).

[1998_12] A. A. Chatzipetros, A. M. Shaarawi, I. M. Besieris, and M. A. Abdel-Rahman, Aperture synthesis of time-limited X waves and analysis of their propagation characteristics, J. Acoust. Soc. Am. **103**, 2287-2295 (1998).

[1998_13] Z. Liu and D. Fan, Propagation of pulsed zeroth-order Bessel beams, J. Mod. Opt. **45**, 17-21 (1998).

[1998_14] I. Garcés, F. Villuendas, J. Subías, J. Alonso, M. del Valle, C. Domínguez, and E. Bartolomé, Bidimensional planar micro-optics for optochemical absorbance sensing, Opt. Lett. **23**, 225-227 (1998).

[1998_15] A. W. Lohmann, D. Mendlovic, and G. Shabtay, Talbot (1836), Montgomery (1967), Lau (1948) and Wolf (1955) on periodicity in optics, Pure Appl. Opt. **7**, 1121-1124 (1998).

[1998_16] F. Mitschke and U. Morgner, The temporal Talbot effect, Optics & Photonics News **9**, 45-47 (1998).

[1998_17] R. Grunwald, R. Ehlert, S. Woggon, and H. H. Witzmann, Mikrolinsen-Arrays für technische Sehsysteme, Project Report, BMBF 01M3025C, April 1998 (in German).

[1998_18] C. Iaconis and I. A. Walmsley, Spectral phase interferometry for direct electric-field reconstruction of ultrashort optical pulses, Opt. Lett. **23**, 792-795 (1998).

[1998_19] M. A. Porras, Ultrashort pulsed Gaussian light beams, Phys. Rev. E **58**, 1086-1093 (1998).

[1999_1] S. Sinzinger and J. Jahns, *Microoptics*, Wiley-VCH, Weinheim, 1999.

[1999_2] R. Grunwald, H. Mischke, and W. Rehak, Microlens formation by thin-film deposition with mesh-shaped masks, Appl. Opt. **38**, 4117-4124 (1999).

[1999_3] N. F. Borrelli, *Microoptics technology - fabrication and applications of lens arrays and devices*, Marcel Dekker, New York, Basel, 1999.

[1999_4] A. V. Kudryashov and H. Weber (Eds.), *Laser resonators - Novel design and development*, SPIE Optical Engineering Press, Bellingham, Washington, 1999.

[1999_5] D. M. Marom, D. Panasenko, P.-Ch. Sun, and Y. Fainman, Spatial-temporal wave mixing for space-time conversion, Opt. Lett. **24**, 563-565 (1999).

[1999_6] R. Grunwald, H.-J. Paetzold, and U. Griebner, Short-Pulse KrF laser with M^2 reduction by a super-Gaussian mirror, Proc. SPIE Vol. **3611**, 102-111 (1999).

[1999_7] U. Griebner and R. Grunwald, Generation of thin film microoptics by crossed deposition through wire grid masks, Proc. SPIE Vol. **3825**, 136-143 (1999).

[1999_8] R. Grunwald, U. Griebner, J. Huschke, and H. Schönnagel, Compact diode-pumped microlaser with mode-selective thin film micro-mirrors, OSA Trends in Optics and Photonics Series (TOPS) on Advanced Solid State Lasers, Vol. **26**, 208-211 (1999).

[1999_9] R. Grunwald, H. Schoennagel, and A. Baerwolff, Frequency conversion of a laser diode MOPA with extra-cavity high-Q resonator and selective feedback, CLEO 1999, Baltimore 1999, paper CWF32, Technical Digest, 270-271.

[1999_10] R. Grunwald, H.-J. Pätzold, and H. Hart, Generation of side-emitting and -detecting fibers with patterned scattering zones by pulse laser treatment, CLEO 1999, Baltimore 1999, paper CWF72, Technical Digest, 298.

[1999_11] C. F. R. Caron and R. M. Potvliege, Optimum conical angle of a Bessel-Gauss beam for low-order harmonic generation in gases, J. Opt. Soc. Am. B **16**, 1377-1384 (1999).

[1999_12] L. Allen, M. J. Padgett, and M. Babiker, The orbital angular momentum of light, Progr. Opt. **39**, 291-372 (1999).

[1999_13] M. Piché, G. Rousseau, C. Varin, and N. McCarthy, Conical wave packets: their propagation speed and their longitudinal fields, Proc. SPIE **3611**, 332-343 (1999).

[1999_14] C. A. McQueen, J. Arlt, and K. Dholakia, An experiment to study a "nondiffracting" light beam, Am. J. Phys. **67**, 912-915 (1999).

[1999_15] Z. Jiang and X.-C. Zhang, 2D-measurement and spatio-temporal coupling of few-cycle THz pulses, Opt. Express **5**, 243-248 (1999).

[1999_16] J.-Y. Lu and S. He, Optical X wave communication, Opt. Commun. **161**, 187-192 (1999).

[1999_17] J. D. Mansell and R. L. Byer, *Sub-lens spatial resolution Shack-Hartmann wavefront sensing*, in: G. D. Love (Ed.), Proc. 2nd international Workshop on Adaptive Optics for Industry and Medicine, World Scientific, Singapore, New Jersey, London, Hong Kong, 1999.

[1999_18] J. D. Mansell, R. L. Byer, and D.R. Neal, *Apodized micro-lenses for Hartmann wavefront sensing*, in: G. D. Love (Ed.), Proc. 2nd international Workshop on Adaptive Optics for Industry and Medicine, World Scientific, Singapore, New Jersey, London, Hong Kong, 1999.

[1999_19] C. Kopp, L. Ravel, and P. Meyrueis, Efficient beamshaper homogenizer design combining diffractive optical elements, microlens array and random phase plate, J. Opt. A: Pure Appl. Opt. **1**, 398-403 (1999).

[1999_20] M. Severi and P. Mottier, Etching selectivity control during resist pattern transfer into silica for the fabrication of microlenses with reduced spherical aberration, Opt. Eng. **38**, 146-150 (1999).

[1999_21] C. Hembd-Sölner, R. F. Stevens, and M. C. Hutley, Imaging properties of the Gabor superlens, J. Opt. A: Pure Appl. Opt. **1**, 94-102 (1999).

[1999_22] Q. Wu, G. D. Feke, and R. D. Grober, Realization of numerical aperture 2.0 using a gallium phosphide solid immersion lens, Appl. Phys. Lett. **75**, 4064-4066 (1999).

[1999_23] A. B. Ruffin, J. V. Rudd, J. F. Whitaker, S. Feng, and H. G. Winful, Direct observations of the Gouy phase shift with single-cycle terahertz pulses, Phys. Rev. Lett. **83**, 3410-3413 (1999).

[1999_24] Z. L. Horváth and Zs. Bor, Reshaping of femtosecond pulses by the Gouy phase shift, Phys. Rev. E **60**, 2337-2346 (1999).

[1999_25] A. A. Babin, A. M. Kiselev, K. I. Pravdenko, A. M. Sergeev, A. N. Stepanov, and E. A. Khazanov, Experimental investigation of the influence of subterawatt femtosecond laser radiation on transparent insulators at axicon focusing, Physics-Uspekhi **42**, 74-77 (1999).

[1999_26] F. R. Caron and R. M. Potvliege, Free-space propagation of ultrashort pulses: space-time couplings in Gaussian pulse beams, J. Mod. Opt. **56**, 1881-1891 (1999).

[1999_27] T. Konishi and Y. Ichioka, Optical spectrogram scope using time-to-two-dimensional space conversion and interferometric time-of-flight cross correlation, Optical Review **6**, 507-512 (1999).

[2000_1] R. Grunwald, Mikrooptische Komponenten für neue Technologien, in: R.-J. Ahlers (Ed.), *Das Handbuch der Bildverarbeitung*, expert-verlag, Renningen-Malmsheim 2000, chapter 2.1.4 (in German).

[2000_2] D. Dragoman, M. Dragoman, and K.-H. Brenner, Optical realization of the ambiguity function of real two-dimensional light sources, Appl. Opt. **39**, 2912-2917 (2000).

[2000_3] R. Grunwald, U. Griebner, F. Tschirschwitz, E. T. J. Nibbering, T. Elsaesser, and V. Kebbel, Femtosecond interference effects generated by refractive thin-film micro-optics, CLEO Europe 2000, Conference Digest, CThJ2, 340.

[2000_4] R. Grunwald, U. Griebner, F. Tschirschwitz, E. T. J. Nibbering, T. Elsaesser,

V. Kebbel, H.-J. Hartmann, and W. Jüptner, Generation of femtosecond Bessel beams with micro-axicon arrays, Optics Letters **25** (2000), 981-983.

[2000_5] S. Nerreter, R. Grunwald, A. Bärwolff, and J. W. Tomm, High-accuracy reflectance mapping of microoptical components, Optical Fabrication and Testing (OFT 2000), OSA Topical Meeting, Quebec, Canada, June 18-22, Post Deadline Paper OPD2-1.

[2000_6] R. Grunwald, U. Griebner, F. Tschirschwitz, E. T. J. Nibbering, P. Hamm, T. Elsaesser, and V. Kebbel, Microoptics for the generation of femtosecond Bessel beam arrays, Diffractive Optics and Micro-Optics (DOMO 2000) OSA Topical Meeting, Quebec, Canada, Paper DTuD31, Technical Digest, 208-210.

[2000_7] N. Matuschek, L. Gallmann, D. H. Sutter, G. Steinmeyer, and U. Keller, Back-side-coated chirped mirrors with ultra-smooth broadband dispersion characteristics, Appl. Phys. B **71**, 509-522 (2000).

[2000_8] J. B. Pendry, Negative refraction makes a perfect lens, Phys. Rev. Lett. **85**, 3966-3669 (2000).

[2000_9] R. Trebino, *Frequency-resolved optical gating: The measurement of ultrashort laser pulses*, Kluwer Academic Publishers, Boston, 2000.

[2000_10] D. Mugnai, A. Rafagni, and R. Ruggeri, Observation of superluminal behaviors in wave propagation, Phys. Rev. Lett. **84**, 4830-4833 (2000).

[2000_11] A. G. Sedukhin, Marginal phase correction of truncated Bessel beams, J. Opt. Soc. Am. A **17**, 1059-1066 (2000).

[2000_12] A. Gürtler, C. Winnewisser, H. Helm, and P. U. Jepsen, Terahertz pulse propagation in the near field and far field, J. Opt. Soc. Am. A **17**, 74-83 (2000).

[2000_13] S. Nerreter, *Entwicklung von Messverfahren für ortsabhängige optische Parameter an laseroptischen Schichtbauelementen*, Diploma Thesis, Technical High School Wildau, 2000 (in German).

[2000_14] T. Tanaka and S. Yamamoto, Comparison of aberration between axicon and lens, Opt. Commun. **184**, 113-118 (2000).

[2000_15] H. Kim, H.-J. Kim, and D.-Y. Park, Partially coherent nondiffracting beams, J. Korean Phys. Soc. **37**, 713-719 (2000).

[2000_16] K. T. McDonald, Bessel beams, arXiv:physics/0006046, 19 Jun. 2000, 1-8.

[2000_17] V. Jarutis, R. Paskauskas, and A. Stabinis, Focusing of Laguerre-Gaussian beams by axicon, Opt. Commun. **184**, 105-112 (2000).

[2000_18] M. A. Porras, R. Borghi, and M. Santarsiero, Few-optical cycle Bessel-Gauss pulsed beams in free space, Phys. Rev. E **62**, 5729-5737 (2000).

[2000_19] J. Salo, J. Fagerholm, A. T. Friberg, and M. M. Salomaa, Unified description of nondiffracting X and Y waves, Phys. Rev. E **62**, 4261-4275 (2000).

[2000_20] J. C. Gutiérrez-Vega, M. D. Iturbe-Castillo, E. Tepichin, G. Ramirez, R. M. Rodriguez-Dagnino, and S. Chávez-Cerda, New member in the family of propagation-invariant optical fields: Mathieu beams, Optics & Photonics News, Dec. 2000, 37-38.

[2000_21] J. C. Gutiérrez-Vega, M. D. Iturbe-Castillo, and S. Chávez-Cerda, Alternative formulation for invariant optical fields: Mathieu beams, Opt. Lett. **25**, 1493-1495 (2000).

[2000_22] D. M. Hartmann, O. Kibar, and S. C. Esener, Characterization of a polymer microlens fabricated by use of the hydrophobic effect, Opt. Lett. **25**, 975-977 (2000).

[2000_23] Microlens is deposited directly onto laser-diode facet, Laser Focus World, Dec. 2000, 13.

[2000_24] I. Hodgkinson, Qi Hong Wu, B. Knight, A. Lakhtakia, and K. Robbe, Vacuum deposition of chiral sculptured thin-films with high optical activity, Appl. Opt. **39**, 642-649 (2000).

[2000_25] F. Yongqi, N. Kok, A. Bryan, N. P. Hung, and O. N. Shing, Experimental study of three-dimensional microfabrication by focused ion beam technology, Rev. Sci. Instrum. **71**, 1006-1008 (2000).

[2000_26] F. Yongqi, N. Kok, and A. Ngoi, Investigation of direct milling of micro-optical elements with continuous relief on a substrate by focused ion beam technology, Opt. Eng. **39**, 3008-3013 (2000).

[2000_27] C. Siegel, F. Loewenthal, J. E. Balmer, and H. P. Weber, Talbot array illuminator for single-shot measurements of laser-induced-damage thresholds of thin-film coatings, Appl. Opt. **39**, 1493-1499 (2000).

[2000_28] R. Szipöcs, A. Kohazi-Kis, S. Lako, P. Apai, A. P. Kovacz, G. DeBell, L. Mott, A. W. Louderback, A. V. Tikhonravov, and M. K. Trubetskov, Negative dispersion mirrors for dispersion control in femtosecond lasers: chirped dielectric mirrors and multi-cavity Gires-Tournois interferometers, Appl. Phys. B **70**, 51-57 (2000).

[2000_29] N. Matuschek, G. Steinmeyer, and U. Keller, Relation between coupled-mode theory and equivalent layers for multilayer interference coatings, Appl. Opt. **39**, 1626-1632 (2000).

[2000_30] J. M. Vigoureux and R. Gust, The use of the hyperbolic plane in studies of multilayers, Opt. Commun. **186**, 231-236 (2000).

[2000_31] F. Villa, Correction masks for thickness uniformity in large-area thin films, Appl. Opt. **39**, 1602-1610 (2000).

[2000_32] R. Diehl (Ed.), *High-power diode lasers - fundamentals, technology, applications*, Springer, Berlin, 2000.

[2000_33] Z. Bouchal and M. Bertolotti, Self-reconstruction of wave packets due to spatio-temporal couplings, J. Mod. Opt. **47**, 1455-1467 (2000).

[2000_34] D. Ding and J. Lu, Higher-order harmonics of limited diffraction Bessel beams, J. Acoust. Soc. Am. **107**, 1212-1214 (2000).

[2000_35] V. N. Belyi, N. S. Kasak, and N. A. Khilo, Frequency conversion of Bessel light beams in nonlinear crystals, Quant. Electron. **30**, 753-766 (2000).

[2000_36] R. M. Herman and T. A. Wiggins, Bessel-like beams modulated by arbitrary radial functions, J. Opt. Soc. Am. A **17**, 1021-1032 (2000).

[2000_37] D. Marom, D. Panasenko, P.-Ch. Sun, and Y. Fainman, Femtosecond-rate space-to-time conversion, J. Opt. Soc. Am. B **17**, 1759-1773 (2000).

[2000_38] D. Marom, D. Panasenko, R. Rokitski, P.-Ch. Sun, and Y. Fainman, Time reversal of ultrafast waveforms by wave mixing of spectrally decomposed waves, Opt. Lett. **25**, 132-134 (2000).

[2000_39] DIN-Taschenbuch 341, Charakterisierung von Laserstrahlen und Laseroptiken - Normen, Beuth Verlag, Berlin, 2000, 30-39 (in German).

[2001_1] R. Grunwald, S. Nerreter, U. Griebner, and H.-J. Kuehn, Design, characterization and application of multilayer micro-optics, Proc. SPIE Vol. **4437**, 40-49 (2001).

[2001_2] J. Baehr, U. W. Krackhardt, and K.-H. Brenner, Fabrication and testing of planar microlens arrays by ion exchange technique in glass, Proc. SPIE Vol. **4455**, 281-292 (2001).

[2001_3] Ph. Nussbaum and H. P. Herzig, Low NA refractive microlenses in fused silica, Optical Engineering **40**, 1412-1414 (2001).

[2001_4] D. Daly, *Microlens arrays*, Taylor & Francis, London, 2001.

[2001_5] N. Karasawa, L. Li, A. Saguro, H. Shigekawa, R. Morita, and M. Yamashita, Optical pulse compression to 5.0 fs by use of only a spatial light modulator for phase compensation, J. Opt. Soc. Am. B **18**, 1742-1746 (2001).

[2001_6] D. M. Hartmann, O. Kinar, and S. C. Esener, Optimization and theoretical modeling of polymer microlens arrays fabricated using the hydrophobic effect, Appl. Opt. **40**, 2736-2746 (2001).

[2001_7] R. Grunwald, U. Griebner, E. T. J. Nibbering, A. Kummrow, M. Rini, T. Elsaesser, V. Kebbel, H.-J. Hartmann, and W. Jüptner, Spatially resolved small-angle non-collinear interferometric autocorrelation of ultrashort pulses with micro-axicon arrays, J. Opt. Soc. Am. A **18** (2001), 2923-2931.

[2001_8] R. Grunwald, U. Griebner, E. T. J. Nibbering, A. Kummrow, T. Elsaesser, and V. Kebbel, Spatial frequency effects of polychromatic ultrashort pulse Bessel beam arrays, CLEO 2001, Baltimore, May 56-11, 2001, Technical Digest, Paper CThS4, 516-517.

[2001_9] R. Grunwald, U. Griebner, T. Elsaesser, V. Kebbel, H.-J. Hartmann, and W. Jüptner, Characterization of the coherence homogeneity of femtosecond lasers by microoptical multichannel interferometry, CLEO/Europe - EQEC Focus Meetings 2001, Progress in Solid State Lasers, Munich, June 18-22, 2001, Conference Digest, 43.

[2001_10] R. Grunwald, U. Griebner, T. Elsaesser, V. Kebbel, H.-J. Hartmann, and W. Jüptner, Femtosecond interference experiments with thin-film micro-optical components, 4th Int. Workshop on Automatic Processing of Fringe Patterns, BIAS Bremen, Sept. 17-19, 2001, in: W. Osten, W. Jüptner (Eds.), *Fringe 2001*, Elsevier, Paris, 2001, 33-40.

[2001_11] R. Grunwald, S. Nerreter, J. W. Tomm, S. Schwirzke-Schaaf, and F. Dörfel, Automated high-accuracy measuring system for specular micro-reflectivity, Proc. SPIE Vol. **4449**, 111-118, (2001).

[2001_12] R. Grunwald, S. Nerreter, U. Griebner, and H.-J. Kuehn, Design, characterization and application of multilayer micro-optics, Proc. SPIE Vol. **4437**, 40-49 (2001).

[2001_13] A. V. Semichaevsky and M. E. Testorf, Anything optical rays cannot do?, Proc. SPIE Vol. **4436**, 56-67 (2001).

[2001_14] T. W. Ng, T. W. Teo, and P. Rajendra, Optical surface roughness evaluation of phosphorus-doped polysilicon films, Optics and Lasers in Engineering **35**, 1-9 (2001).

[2001_15] A. N. Khilo, E. G. Katranji, and A. A. Ryzhevich, Axicon-based Bessel resonator: analytical description and experiment, J. Opt. Soc. Am. A **18**, 1986-1992 (2001).

[2001_16] R. Piestun and D. A. B. Miller, Spatiotemporal control of ultrashort optical pulses by refractive-diffractive-dispersive structured optical elements, Opt. Lett. **26**, 1373-1375 (2001).

[2001_17] R. M. Herman and T. A. Wiggins, Propagation and focusing of Bessel-Gauss, generalized Bessel-Gauss, and modified Bessel-Gauss beams, J. Opt. Soc. Am. A **18**, 170-176 (2001).

[2001_18] J. Salo and M. M. Salomaa, Diffraction-free pulses at arbitrary speeds, J. Opt. A: Pure Appl. Opt. **3**, 366-373 (2001).

[2001_19] J. Salo, A. Friberg, and M. M. Salomaa, Orthogonal X waves, J. Phys. A: Math. Gen. **34**, 9319-9327 (2001).

[2001_20] J. C. Gutiérrez-Vega, M. D. Iturbe-Castillo, E. Tepichin, G. Ramirez, R. M. Rodriguez-Dagnino, S. Chávez-Cerda, and G. H. C. New, Experimental demonstration of optical Mathieu beams, Opt. Commun. **195**, 35-40 (2001).

[2001_21] E.-B. Kley, A. Tünnermann, U. D. Zeitner, and W. Karthe, Micro-optical elements, functions and lithographic fabrication technologies, LaserOpto **33**, 78-84 (2001).

[2001_22] A. Schilling, H. P. Herzig, L. Stauffer, U. Vokinger, and M. Rossi, Efficient beam shaping of linear, high-power diode-lasers using micro-optics, Appl. Opt. **40** (32), 5852-5859, (2001).

[2001_23] T. Hessler and D. Daly, Refractive Mikrolinsenraster besitzen universelle Anwendungsmöglichkeiten, Photonik No. 2, 2-4 (2001).

[2001_24] P. Xi, C. Zhou, S. Zhao, E. Dai, and L. Liu, Pulse-width measurement of ultrashort laser pulse based on Talbot effect, Proc. SPIE Vol. **4438**, 116-123 (2001).

[2001_25] J. Jahns, E. ElJoudi, D. D. Hagedorn, and S. Kinne, Talbot interferometer as a time filter, Optik **112**, 295-298 (2001).

[2001_26] J. Tervo and J. Turunen, Self-imaging of electromagnetic fields, Opt. Express **9**, 622-630 (2001).

[2001_27] M. T. Flores-Arias, M. V. Pérez, C. Gómez-Reino, and C. Bao, Talbot effect reinterpreted by number theory, J. Opt. Soc. Am. A **18**, 2797-2716 (2001).

[2001_28] J. A. Davis, D. E. McNamara, and D. M. Cottrell, Encoding complex diffractive optical elements onto a phase-only liquid-crystal spatial light modulator, Opt. Eng. **40**, 327-329 (2001).

[2001_29] D. Fletcher, K. B. Crozier, K. W. Guarini, S. C. Minne, G. S. Kino, C. F. Quate, and K. E. Goodson, Microfabricated silicon solid immersion lens, J. Micromech. Systems 10, 450-459 (2001).

[2001_30] F. C. DeSchryver, *Femtochemistry. With the Nobel lecture of A. Zewail*, Wiley-VCH, Weinheim, 2001.

[2001_31] R. Menzel, *Photonics - Linear and nonlinear interactions of laser light and matter*, Springer, Berlin, 2001.

[2001_32] P. Vucusic, R. Sambles, C. Lawrence, and G. Wakely, Sculptured-multilayer optical effects in two species of Papilio butterfly, Appl. Opt. **40**, 1116-1125 (2001).

[2001_33] L. L. Sánchez-Soto, J. J. Monzón, T. Yonte, and J. F. Cariñena, Simple trace criterion for classification of multilayers, Opt. Lett. **26**, 1400-1402 (2001).

[2001_34] M. Mansuripur, Omni-directional dielectric mirrors, Optics & Photonics News, Sept. 2001, pp. 46-50.

[2001_35] Z. Bouchal, Self-reconstruction ability of wave field, Proc. SPIE Vol. **4356**, 217-224 (2001).

[2001_36] S. Feng and H. G. Winful, Physical origin of the Gouy phase shift, Opt. Lett. **26**, 485-487 (2001).

[2001_37] A. Marcinkevicius, S. Juodkazis, V. Mizeikis, S. Matsuo, and H. Misawa, Application of femtosecond Bessel-Gauss beam in microstructuring of transparent materials, Proc. SPIE Vol. **4271**, 150-158 (2001).

[2001_38] P. O'Shea, M. Kimmel, X. Gu, and R. Trebino, Highly simplified device for ultrashort-pulse measurement, Opt. Lett. **26**, 932-934 (2001).

[2001_39] I. Walmsley, L. Waxer, and C. Dorrer, The role of dispersion in ultrafast optics, Rev. Sci. Instrum. **72**, 1-29 (2001).

[2001_40] L. Gallmann, G. Steinmeyer, D. H. Sutter, T. Rupp, C. Iaconis, I. A. Walmsley, and U. Keller, Spatially resolved amplitude and phase characterization of femtosecond optical pulses, Opt. Lett. **26**, 96-98 (2001).

[2001_41] R. Trebino, P. O'Shea, M. Kimmel, and X. Gu, Measuring ultrashort laser pulses - just got a lot easier, Optics & Photonics News, June 2001, 22-25.

[2002_1]   L. van Beurden, M:cube eases coupling problems, FibreSystems Europe, Feb. 2002, p. 13; http://fibers.org/articles/news/4/2/18/1.

[2002_2]   T. H. Oakley and C. W. Cunningham, Molecular phylogenetic evidence for the independent evolutionary origin of an arthropod compound eye, Proc. Natl. Acad, Sci. USA, Vol. 99, Issue 3, 1426-1430, Feb. 5, 2002.

[2002_3]   R. Morita, M. Hirasawa, N. Karasawa, S. Kusaka, N. Nakagawa, K. Yamane, L. Li, A. Suguro, and M. Yamashita, Sub-5-fs optical pulse characterization, Meas. Sci. Technol. **13**, 1710-1720 (2002).

[2002_4]   G. Seewald and R. Grunwald, Spatially resolved measurement of slow-axis pseudo near-field of diode laser arrays, Proc. SPIE Vol. **4833**, 900-905 (2002).

[2002_5]   R. Grunwald, U. Griebner, U. Neumann, A. Kummrow, E. T. J. Nibbering, M. Rini, V. Kebbel, M. Piché, G. Rousseau, and M. Fortin, Femtosecond laser beam shaping with structured thin-film elements, Proc. SPIE Vol. **4833**, 354-361 (2002).

[2002_6]   R. Grunwald, U. J. Neumann, and V. Kebbel, Femtosecond laser wavefront sensing with thin-film microaxicon arrays, OSA Annual Meeting, Orlando, Sept. 29 – Oct. 03, 2002, Conference Program, 129, paper ThQ5.

[2002_7]   R. Grunwald, U. Griebner, U. Neumann, A. Kummrow, E. T. J. Nibbering, M. Piché, G. Rousseau, M. Fortin, and V. Kebbel, Generation of ultrashort-pulse nondiffracting beams and X-waves with thin-film axicons, 13th Int. Conf. on Ultrafast Phenomena, May 12-17, 2002, Vancouver, Canada, Technical Digest, 70-71, and in: M. Murnane, N. F. Scherer, and A. M. Weiner (Eds.), *Ultrafast Phenomena XIII*, Springer-Verlag, New York, 2002, 247-249.

[2002_8]   F. Dörfel, S. Nerreter, J. W. Tomm, R. Grunwald, R. Kunkel, and J. Luft, Micro-Reflectance Inspection of Diode Laser Front Facets, Proc. SPIE Vol. **4648** (2002), 48-54.

[2002_9]   R. Grunwald, V. Kebbel, U. Griebner, E. T. J. Nibbering, A. Kummrow, M. Rini, T. Elsaesser, and W. Jüptner, Spectral shaping of femtosecond X-waves, CLEO 2002, May 19-24, Long Beach, CA, Technical Digest, 422-423.

[2002_10] R. Grunwald, V. Kebbel, U. Griebner, E. T. J. Nibbering, A. Kummrow, and M. Rini, Coherence mapping of femtosecond lasers, CLEO 2002, May 19-24, Long Beach, CA, Technical Digest, 84-85.

[2002_11] P. Gori and M. Pappalardo, Propagation-independent fields, Ultrasonics **40**, 287-291 (2002).

[2002_12] W. Hu and H. Guo, Ultrashort pulsed Bessel beams and spatially induced group-velocity dispersion, J. Opt. Soc. Am. **A 19**, 49-53 (2002).

[2002_13] K. Volke-Sepulveda, V. Garcés, S. Chávez-Cerda, J. Arlt, and K. Dholakia,

Orbital angular momentum of a high-order Bessel light beam, J. Opt. B: Quantum Semiclass. Opt. **4**, S82-S89 (2002).

[2002_14] E. Hasman, Z. Bomzon, A. Niv, G. Biener, and V. Kleiner, Polarization beam-splitters and optical switches based on space-variant computer-generated subwavelength quasi-periodic structures, Opt. Commun. **209**, 45-54 (2002).

[2002_15] B. Dépret, P. Verkerk, and D. Hennequin, Characterization and modelling of the hollow beam produced by a real conical lens, Opt. Commun. **211**, 31-38 (2002).

[2002_16] A. Fischer, *Entwicklung von Messverfahren zur winkel- und ortsaufgelösten Reflexionsmessung an Multilayer-Gradienten-Mikrooptiken*, Diploma Thesis, Technical High School Wildau, 2002 (in German).

[2002_17] G. Seewald, *Charakterisierung der Strahleigenschaften von Diodenlasern hinsichtlich Divergenz und Polarisation*, Diploma Thesis, Technical High School Wildau, 2002 (in German).

[2002_18] J.-F. Fortin, G. Rousseau, N. McCarthy, and M. Piché, Generation of quasi-Bessel beams and femtosecond optical X-waves with conical mirrors, Proc. SPIE Vol. **4833**, 876-884 (2002).

[2002_19] S. R. Seshadri, Virtual source for the Bessel-Gauss beam, Opt. Lett. **27**, 998-1000 (2002).

[2002_20] E. Comay, Remarks on the physcial meaning of diffraction-free solutions of the Maxwell equations, arXiv:physics/0202030v1, 10 Feb. 2002.

[2002_21] S. Chávez-Cerda, D. Iturbe-Castillo, and J. C. Gutiérrez-Vega, Elliptic propagation invariant optical fields: Mathieu beams, Proc. SPIE Vol. **4833**, 369-377 (2002).

[2002_22] S. Chávez-Cerda, M. J. Padgett, I. Allison, G. H. C. New, J. C. Gutiérrez-Vega, A. T. O'Neil, I. MacVicar, and J. Courtila, Holographic generation and orbital angular momentum of high-order Mathieu beams, J. Opt. B: Quantum Semiclass. Opt. **4**, S52-S57 (2002).

[2002_23] X.-C. Yuan, W. X. Yu, N. Q. Ngo, and W. C. Cheong, Cost-effective fabrication of microlenses on hybrid sol-gel glass with a high-energy beam-sensitive gray-scale mask, Opt. Express **10**, 303-308 (2002).

[2002_24] Q. Peng, Y. Guo, S. Liu, and Z. Cui, Real-time gray-scale photolithography for fabrication of continuous microstructure, Opt. Lett. **27**, 1720-1722 (2002).

[2002_25] Y. Fu and N. K. A. Bryan, Inverstigation of hybrid microlens integration with vertical-cavity surface-emitting lasers for free-space optical links, Opt. Express **10**, 413-418 (2002).

[2002_26] Z. Bouchal, Controlled spatial shaping of nondiffracting patterns and arrays, Opt. Lett. **27**, 1376-1378 (2002).

[2002_27] I. Alexeev, K. Y. Kim, and H. M. Milchberg, Measurement of the superluminal group velocity of an ultrashort Bessel beam pulse, Phys. Rev. Lett. **88**, 073901-1 (2002).

[2002_28] P. Xi, C. Zhou, E. Dai, and L. Liu, Novel method for ultrashort laser pulse-width measurement based on the self-diffraction effect, Opt. Express **10**, 1099-1104 (2002).

[2002_29] C. Dorrer, Comment on: Novel method for ultrashort laser pulse-width measurement based on the self-diffraction effect, Opt. Express **11**, 79-80 (2002).

[2002_30] Z. Bomzon, A. Niv, G. Biener, V. Kleiner, and E. Hasman, Polarization Talbot self-imaging with computer-generated, space variant subwavelength dielectric gratings, Appl. Opt. **41**, 5218-5222 (2002).

[2002_31] P. Xi, C. Zhou, E. Dai, and L. Liu, New pulse-width measurement for ultrashort laser pulse, OSA TOPS **75**, 45-51 (2002).

[2002_32] N. Lindlein, Simulation of micro-optical systems including microlens arrays, J. Opt. A: Pure Appl. Opt. **4**, 1-9 (2002).

[2002_33] S. E. Lyshevski, *MEMS and NEMS: Systems, Devices and Structures*, CRC Press, Boca Raton, FL, 2002.

[2002_34] V. Saile, L. Shabel'nikov, V. Nazmov, F. J. Pantenburg, J. Mohr, V. Yunkin, S. Kouznetsov, V. Pindyurin, I. Snigirewa, and A. Snigirev, X-ray lens with kinoform refractive profile created by X-ray lithography, Proc. SPIE Vol. **4783**, 176-184 (2002).

[2002_35] T. Konishi and Y. Ichioka, Ultrafast temporal-spatial optical information processing, conversion, and transmission, in: H. J. Caulfield (Ed.), *Optical information processing: A tribute to Adolf Lohmann*, SPIE Press, Bellingham, Washington, 2002, 311-356.

[2002_36] R. R. Willey, *Practical design and production of optical thin-films*, Marcel Dekker, New York, 2002.

[2002_37] H. A. Macleod, *Thin-film optical filters*, IoP Institute of Physics Publishing, Bristol, England, 2002 (3nd Edition).

[2002_38] M. Kölbel, R. W. Tjerkstra, J. Brugger, C. J. M. van Rijn, W. Nijdam, J. Huskens, and D. N. Reinhoudt, Shadow-mask evaporation through monolayer-modified nanostencils, Nano Lett. **2**, 1339-1343 (2002).

[2002_39] K. D. Möller, *Optics - learning by computing, with examples using MATHCAD®*, Springer, New York, 2002.

[2002_40] R. Giust and J.-M. Vigoureux, Hyperbolic representation of light propagation in a multilayer medium, J. Opt. Soc. Am. A **19**, 378-384 (2002).

[2002_41] J. I. Larruquert, New layer-by-layer multilayer design method, J. Opt. Soc. Am. A **19**, 385-390 (2002).

[2002_42] V. Kebbel, J. Mueller and W. P. O. Jueptner, Characterization of aspherical micro-optics using digital holography: improvement of accuracy, Proc. SPIE Vol. **4778**, 188-197 (2002).

[2002_43] K. Reivelt and P. Saari, Experimental demonstration of realizability of optical focus wave modes Phys. Rev. E **66**, 056611 (2002).

[2002_44] V. Garcés-Chávez, D. McGloin, H. Melville, W. Sibbett, and K. Dholakia, Simultaneous Micromanipulation in multiple planes using a self-reconstructing light beam, Nature **419**, 145-147 (2002).

[2002_45] Z. Bouchal, Resistance of nondiffracting vortex beam against amplitude and perturbations, Opt. Commun. **210**, 155-164 (2002).

[2002_46] D.-S. Ding, J.-Y. Xu, and Y.-J. Wang, Second-harmonic generation of Bessel beams in lossy media, Chin. Phys. Lett. **19**, 689-690 (2002).

[2002_47] M. Nisoli, E. Priori, G. Sansone, S. Stagira, G. Cerullo, S. De Silvestri, C. Altucci, R. Bruzzese, C. de Lisio, P. Villoresi, L. Polctto, M. Pascolini, and G. Tondello, High brightness high-order harmonic generation by truncated Bessel beams in the Sub-10-fs regime, Phys. Rev. Lett. **88**, 033902-1 to 033902-4 (2002).

[2002_48] C. Dorrer and A. N. Walmsley, Linear measurement of space-time coupling in ultrashort optical pulses and definition of a degree of spatio-temporal uniformity, CLEO '02, Technical Digest, 346-347 (2002).

[2002_49] C. Dorrer and I. Walmsley, Simple linear technique for the measurement of space-time coupling in ultrashort optical pulses, Opt. Lett. **21**, 1947-1949 (2002).

[2002_50] C. Dorrer, E. M. Kosik, and I. A. Walmsley, Direct space-time characterization of the electric fields of ultrashort optical pulses, Opt. Lett. **27**, 548-550 (2002).

[2003_1] R. Grunwald, U. Neumann, U. Griebner, K. Reimann, G. Steinmeyer, and V. Kebbel, Ultrashort-pulse wavefront autocorrelation, Opt. Lett. **28**, 2399-2401 (2003).

[2003_2] D. Dudley, W. M. Duncan, and J. Slaughter, Emerging digital mirror device (DMD) applications, Proc. SPIE Vol. **4985**, 14-15 (2003).

[2003_3] R. Völkel, M. Eisner, and K. J. Weible, Miniaturized imaging systems, Microelectronic Engineering, **461-472**, 67-68 (2003).

[2003_4] Y. S, Kivshar and G. P. Agrawal, *Optical solitons - From fibers to photonic crystals*, Academic Press, Elsevier Science, Amsterdam, 2003.

[2003_5] G. Brooker, *Modern Classical Optics*, Oxford University Press, Oxford, 2003.

[2003_6] A. Thaning, Z. Jaroszewicz and A. T. Friberg, Diffractive axicons in oblique illumination: analysis and experiments and comparison with elliptical axicons, Appl. Opt. **42**, 9-17 (2003).

[2003_7] R. Grunwald, V. Kebbel, U. Griebner, U. Neumann, A. Kummrow, M. Rini,

E.T.J. Nibbering, M. Piché, G. Rousseau, and M. Fortin, Generation and characterization of spatially and temporally localized few-cycle optical wavepackets, Phys. Rev. A **67**, 063820 (2003); co-published in July 2003 issue of Virtual Journal of Ultrafast Science (http://www.vjultrafast.org).

[2003_8] R. Grunwald, V. Kebbel, U. Neumann, U. Griebner, and M. Piché, Spatio-temporal processing of femtosecond laser pulses with thin-film microoptics, Proc. SPIE Vol. **5181**, 1-11 (2003).

[2003_9] R. Grunwald, U. Neumann, V. Kebbel, and K. Mann, Thin-film microaxicon arrays for beam shaping and characterization of vacuum ultraviolet lasers, CLEO Europe 2003, Munich, June 22-27, 2003, Conference Digest (CD-ROM), CM3-2-THU.

[2003_10] U. Neumann, R. Grunwald, U. Griebner, G. Steinmeyer, M. Woerner, and W. Seeber: Second harmonic characteristics of photonic composite glass layers with ZnO nanocrystallites for ultrafast applications, Proc. SPIE Vol. **4972**, 112-121 (2003).

[2003_11] R. Grunwald, U. Neumann, U. Griebner, M. Rini, E.T.J. Nibbering, A. Kummrow, V. Kebbel, M. Piché, G. Rousseau, and M. Fortin, Spatio-temporal autocorrelation of ultrashort-pulse localized wavepackets, CLEO/QELS 2003, Baltimore, June 1-6, 2003, Technical Digest (CD-ROM), Paper QTuG25.

[2003_12] A. J. Abu El-Haija, Effective medium approximation for the effective optical constants of a bilayer and a multilayer structure based on the characteristic matrix technique, J. Appl. Phys. **93**, 2590-2594 (2003).

[2003_13] E. Recami, M. Zamboni-Rached, K. Z. Nóbrega, C. A. Dartora, and H. E. Hernández F., On the localized superluminal solutions to the Maxwell equations, IEEE J. Selected Topics Quant. Electron. **9**, 59-73 (2003).

[2003_14] M. de Angelis, L. Cacciapouti, G. Pierattini, and G. M. Tino, Axially symmetric hollow beams using refractive conical lenses, Opt. Laser Eng. **39**, 283-291 (2003).

[2003_15] V. Vaicaitis and S. Paulikas, Formation of Bessel beams with continuously variable cone angle, Optical and Quantum Electronics **35**, 1065-1071 (2003).

[2003_16] F. Di, Y. Yingbai, J. Guofan, H. Liu, and T. Qiaofeng, Iterative optimization algorithm based on electromagnetic model for designing diffractive micro-optical elements, Opt. Commun. **220**, 7-15 (2003).

[2003_17] C. Quan, S. H. Wang, C. J. Tay, I. Reading, and Z. P. Fang, Integrated optical inspection on surface geometry and refractive index distribution of a microlens array, Opt. Commun. **225**, 223-231 (2003).

[2003_18] W. X. Yu and X. C. Yuan, Fabrication of refractive microlens in hybrid SiO$_2$/TiO$_2$ sol-gel glass by electron beam lithography, Opt. Express **11**, 899-903 (2003).

[2003_19] M. Karlsson and F. Nikolajeff, Diamond micro-optics: microlenses and antireflection structured surfaces for the infrared spectral region, Optics Express **11**, 502-507 (2003).

[2003_20] S. Minardi, A. Varanavicus, A. Piskarskas, and P. Di Trapani, A compact multi-pixel parametric light source, Opt. Commun. **224**, 301-307 (2003).

[2003_21] P. Di Trapani, G. Valiulis, A. Piskarskas, O. Jedrkiewicz, J. Trull, C. Conti, and S. Trillo, Spontaneously generated X-shaped light bullets, Phys. Rev. Lett. **91**, 093904-1 (2003).

[2003_22] N. Guérineau, E. Di Mambro, and J. Primot, Talbot experiment re-examined: study of the chromatic regime and application to spectrometry, Opt. Express **11**, 3310-3319 (2003).

[2003_23] J. Jahns, H. Knuppertz, and A. Lohmann, Montgomery self-imaging effect using computer-generated diffractive optical elements, Opt. Commun. **225**, 13-17 (2003).

[2003_24] J. Békési, D. Schäfer, J. Ihlemann, and P. Simon, Fabrication of diffractive optical elements by ArF-laser ablation of fused silica, Proc. SPIE Vol. **4977**, 235-240 (2003).

[2003_25] V. Shaoulov and J. P. Rolland, Design and assessment of microlenslet-array relay optics, Appl. Opt. **42**, 6838–6845 (2003).

[2003_26] N. Kaiser and H. K. Pulker (Eds.), *Optical Interference Coatings*, Springer, Berlin, 2003.

[2003_27] L. Allen, S. M. Barnett, and M. J. Padgett, *Optical Angular Momentum*, Institute of Physics Publishing, Bristol, 2003.

[2003_28] V. Messager, F. Louradour, C. Froehly, and A. Barthelemy, Coherent measurement of short laser pulses based on spectral interferometry resolved in time, Opt. Lett. **28**, 743-745 (2003).

[2003_29] I. D. Nikolov, K. Kurihara, and K. Goto, Nanofocusing probe optimization with anti-reflection coatings for a high-density optical memory, Nanotechnology **14**, 946-954 (2003).

[2003_30] P. Saari, Relativistic Doppler effect, aberration and Gouy effect on localized waves, Atti della Fondazione Giorgio Ronchi, Anno LVIII, n. 6, 729-754 (2003).

[2003_31] D. J. Biaswas, J. P. Nilaya and M. B. Danailov, Enhancement of photo-refractive two-wave mixing gain with a Bessel pump beam, Opt. Commun. **226**, 387-391 (2003).

[2003_32] C. Altucci, R. Bruzzese, C. de Lisio, M. Nisoli, E. Priori, S. Stagira, M. Pascolini, L. Poletto, P. Villoresi, V. Tosa and K. Midorikawa, Phase-matching analysis of high-order harmonics generated by truncated Bessel beams in the sub-10-fs regime, Phys. Rev. A **68**, 033806 (2003).

[2003_33] S. Akturk, M. Kimmel, P. O'Shea, and R. Trebino, Measuring spatial chirp in ultrashort pulses using single-shot frequency-resolved optical gating, Opt. Expr. **11**, 68-78 (2003).

[2003_34] S. Akturk, M. Kimmel, P. O'Shea, and R. Trebino, Measuring pulse front tilt in ultrashort pulses using GRENOUILLE, Opt. Expr. **11**, 491-501 (2003).

[2004_1] R. Grunwald, U. Neumann, V. Kebbel, H.-J. Kühn, K. Mann, U. Leinhos, H. Mischke, and D. Wulff-Molder, VUV beam array generation by flat microoptical structures, Opt. Lett. **29**, 977-979 (2004).

[2004_2] R. Grunwald, U. Neumann, U. Griebner, V. Kebbel, and H.-J. Kuehn, Spatio-temporal control of laser beams with thin-film shapers, Proc. SPIE Vol. **5333**, 1-11 (2004).

[2004_3] J. Jang, Y.-S. Oh, and B. Javidi, Spatiotemporally multiplexed integral imaging projector for large-scale high-resolution three-dimensional display, Opt. Express **12**, 557-563 (2004).

[2004_4] J. Jahns and K.-H. Brenner (Eds.) *Microoptics - from technology to applications*, Springer Series in Optical Sciences, Springer Verlag, 2004.

[2004_5] R. Grunwald, V. Kebbel, U. Neumann, U. Griebner, and M. Piché, Ultrafast spatio-temporal processing with thin-film microoptics, Opt. Eng. **43**, 2518-2524 (2004).

[2004_7] U. Neumann, R. Grunwald, U. Griebner, G. Steinmeyer, and W. Seeber, Second harmonic efficiency of ZnO nanolayers, Appl. Phys. Lett. **84**, 170-172 (2004).

[2004_8] R. Grunwald and G. Steinmeyer, Ultrakurze Lichtblitze: Regenbögen in Raum und Zeit, Physik in unserer Zeit **35**, 218-226 (2004) (in German).

[2004_9] R. Grunwald, U. Neumann, and V. Kebbel, Generation of small-angle localized light waves with thin-film microoptics, Progress in Electromagnetics Research Symposium (PIERS) 2004, Workshop on Localized waves, Pisa, Italy, March 28-31, 2004, Conference Proceedings (CD-ROM), Session 61, paper 8.

[2004_10] P. Gabolde and R. Trebino, Self-referenced measurement of the complete electric field of ultrashort pulses, Opt. Express **19**, 4423-4429 (2004).

[2004_11] R. Grunwald, U. Neumann, G. Stibenz, S. Langer, G. Steinmeyer, V. Kebbel, J.-L. Néron, and M. Piché, Truncated ultrashort-pulse small-angle Bessel beams, Proc. SPIE Vol. **5579**, 724-735 (2004).

[2004_12] R. Grunwald, U. Griebner, U. Neumann, and V. Kebbel, Self-reconstruction of ultrashort-pulse Bessel-like X-waves, CLEO/QELS 2004, San Francisco, May 16-21, 2004, Conference Digest (CD-ROM), paper number CMQ7.

[2004_13] R. Grunwald, U. Neumann, U. Griebner, K. Reimann, G. Steinmeyer, and V. Kebbel, Wavefront autocorrelation of femtosecond laser beams, Proc. SPIE Vol. **5333**, 122-130 (2004).

[2004_14] S. Langer, R. Grunwald, U. Neumann, and V. Kebbel, Tilt-tolerance of microaxicons for wavefront sensing of highly divergent beams, 10th Microoptics Conference MOC 2004, Sept. 1-3, 2004, Jena, Proceedings CD-ROM, paper F20.

[2004_15] R. Grunwald, U. Neumann, V. Kebbel, and H.-J. Kühn, Microoptical beam shaping of ultrashort laser pulses with low-dispersion components, 10th Microoptics Conference MOC 2004, Sept. 1-3, 2004, Jena, Proceedings CD-ROM, paper F21.

[2004_16] M. Wegener, *Extreme Nonlinear Optics*, Springer-Verlag, Berlin 2005.

[2004_17] S. V. Kukhlevsky and M. Mechler, Diffraction-free subwavelength-beam optics at nanometer scale, Opt. Commun. **231**, 35-43 (2004).

[2004_18] P. Saari and K. Reivelt, Generation and classification of localized waves by Lorentz transformations in Fourier space, Phys. Rev. E **69**, 036612 (2004).

[2004_19] S. Langer, *Neuartiges 2D-Autokorrelatorsystem für ultrakurze Laserimpulse basierend auf einem erweiterten Shack-Hartmann Verfahren*, Master Thesis, Technical High School Wildau, 2004 (in German).

[2004_20] I. M. Besieris, A. M. Shaarawi, Paraxial localized waves in free space, Opt. Express **12**, 3848-3864 (2004).

[2004_21] V. Kebbel, *Untersuchungen zur Erzeugung und Propagation ultrakurzer optischer Bessel-Impulse*, Doctoral Thesis, University Bremen, 2004, in: F. Vollertsen, W. Jüptner (Eds.), Strahltechnik, Band **25**, Bremer Institut für angewandte Strahltechnik, Bremen, 2004 (in German).

[2004_22] J. Azaña, N. K. Berger, B. Levit, V. Smulakovsky, and B. Fischer, Frequency shifting of microwave signals by use of a general temporal self-imaging (Talbot) effect in optical fibers, Opt. Lett. **29**, 2849-2851 (2004).

[2004_23] C. Zhou, W. Wang, E. Dai, and L. Liu, Simple principles of the Talbot effect, Optics & Photonics News, Nov. 2004, 46-50.

[2004_24] G. Mínguez-Vega and J. Jahns, Temporal processing with the Montgomery interferometer, Opt. Commun. **236**, 45-52 (2004).

[2004_25] J. Duparré, P. Schreiber, P. Dannberg, T. Scharf, P. Pelli, R. Völkel, H.-P. Herzig, and A. Bräuer, Artifical compound eyes - different concepts and their application to ultra flat image acquisition sensors, Proc. SPIE Vol. **5346**, 89–100 (2004).

[2004_26] J. Duparré, P. Dannberg, P. Schreiber, A. Bräuer, and A. Tünnermann, Artificial apposition compound eye fabricated by micro–optics technology, Appl. Opt. **43**, 4303–4310 (2004).

[2004_27] A. Kosiorek, W. Kandulski, P. Chudzinski, K. Kempa, and M. Giersig, Shadow nanosphere lithography: simulation and experiment, Nano Lett. **4**, 1359-1363 (2004).

[2004_28] M. Roy, I. Cooper, P. Moore, C. J. R. Sheppard, and P. Hariharan, White-light interference microscopy: effects of multiple reflections within a surface film, Opt. Express **13**, 164-170 (2004).

[2004_29] S. Varró, Scattering of a few-cycle laser pulse on a thin metal layer: the effect of the carrier-envelope phase difference, Laser Phys. Lett. **1**, 42-45 (2004).

[2004_30] R. Grunwald, S. Nerreter, G. Seewald, A. Fischer, U. Neumann, and M. Tischer, Mikrooptische Dünnschicht-Komponenten zur Effizienzverbesserung der Kollimation von Diodenlaser-Arrays, in: *Mikrooptiken für Hochleistungsdiodenlaser*, VDI-Technologiezentrum Düsseldorf, 2004, 65-79 (in German).

[2004_31] K. Reivelt and P. Saari, Bessel-Gauss pulse as an appropriate mathematical model for optically realizable localized waves, Opt. Lett. **29**, 1176-1178 (2004).

[2005_1] A. Lakhtakia and R. Messier, *Sculptured Thin Films - Nanoengineered morphology and optics*, SPIE Press, Bellingham, Washington, 2005.

[2005_2] M. Mansuripur, *Classical optics and its applications*, Cambridge University Press, Cambridge, 2005.

[2005_3] Q. Yang, L. Liu, and H. Lang, Computation of the ambiguity function for circularly symmetric pupils, J. Opt. A: Pure Appl. Opt. **7**, 431-437 (2005).

[2005_4] W. C. Cheong, B. P. S. Ahluwalia, X.-C. Yuan, L.-S. Zhang, H. Wang, H. B. Niu, and X. Peng, Fabrication of efficient microaxicon by direct electron-beam lithography for long nondiffracting distance of Bessel beams for optical manipulation, Appl. Phys. Lett. **87**, 024104 (2005).

[2005_5] R. Grunwald, Microoptics, in: G. L. Trigg (Ed.), *Digital Encyclopedia of Applied Physics*, Wiley-VCH, Weinheim, Edition 2005, http://www.mrw.interscience.wiley.com/eap/articles/eap647/frame.html.

[2005_6] R. Grunwald, Mikrooptiken für extreme Laserparameter, Laser Technik Journal **2**, No. 3, 51-55 (2005) (in German).

[2005_7] R. Grunwald and V. Kebbel, Ultrafast and ultrabroadband spatio-temporal processing with thin-film microoptics, Frontiers in Optics 2005, 89th OSA Annual Meeting, Topical Meeting on Passive Photonic Devices, Tucson, Arizona, October 17-20, 2005, Conference Proceedings (CD-ROM), Paper FWS4.

[2005_8] R. Grunwald, U. Neumann, U. Griebner, G. Stibenz, G. Steinmeyer, V. Kebbel, and M. Piché, Recent advances in thin-film microoptics, Proc. SPIE Vol. **5827,** 187-199 (2005).

[2005_9] U. Neumann, R. Grunwald, A. Rosenfeld, J. Li, and P. Herman, Deep-UV laser microstructuring of dielectric materials with arrays of nondiffracting beams, CLEO Europe 2005, Munich, June 22-27, 2005, Conference Digest (CD-ROM), Paper CM1-3-TUE.

[2005_10] R. Grunwald, U. Neumann, and H.-J. Kühn, Nanolayer microoptics for laser beam shaping at extreme parameters, CLEO Europe 2005, Munich, June 22-27, 2005, Paper CK2-3-THU.

[2005_11] R. Grunwald, S. Langer, U. Neumann, U. Griebner, G. Stibenz, G. Steinmeyer, V. Kebbel, M. Piché, and J.-L. Néron, Characterization of ultrashort pulses with spatio-temporal and angular resolution, CLEO 2005,

Baltimore, May 23-27 (2005), Conference Digest (CD-ROM), paper CThP2.

[2005_12] R. Grunwald, U. Neumann, S. Langer, G. Stibenz, G. Steinmeyer, V. Kebbel, M. Piché, and J.-L. Néron, Ultrabroadband spectral transfer in extended focal zones: truncated few-cycle Bessel-Gauss beams, CLEO 2005, Baltimore, May 23-27 (2005), Conference Digest (CD-ROM), paper CTuAA6.

[2005_13] U. Neumann, R. Grunwald, U. Griebner, G. Steinmeyer, and W. Seeber, High second harmonic conversion efficiency of ZnO nanolayers for ultrafast laser applications, CLEO Europe 2005, Munich, June 22-27, 2005, Technical Digest CD-ROM, Poster CD-22-MON.

[2005_14] M. Sheik-Bahae, Nonlinear Optics Basics. Kramers-Kronig relations in nonlinear optics, in: Robert D. Guenther (Ed.): *Encyclopedia of Modern Optics*, Academic Press, Amsterdam 2005.

[2005_15] M. Zamboni-Rached, E. Recami, and H. E. Hernández-Figueroa, Theory of "frozen waves": modeling the shape of stationary wave fields, J. Opt. Soc. Am. A **22**, 2465-2475 (2005).

[2005_16] R. I. Hernández-Aranda, S. Chávez-Cerda, and J. C. Gutiérrez-Vega, Theory of the unstable Bessel resonator, J. Opt. Soc. Am. A **22**, 1909-1917 (2005).

[2005_17] P. Grelu, J. M. Soto-Crespo, and N. Akhmediev, Light bullets and dynamic pattern formation in nonlinear dissipative systems, Opt. Express **13**, 9352-9360 (2005).

[2005_18] A. Chafiq, Z. Hricha, and A. Belafhal, Paraxial propagation of Mathieu beams through an apertured ABCD optical system, Opt. Commun. **253**, 223-230 (2005).

[2005_19] D. McGloin and K. Dholakia, Bessel beams: diffraction in a new light, Contemporary Physics **46**, 15-28 (2005).

[2005_20] P. Fischer, C. T. A. Brown, J. E. Morris, C. López-Mariscal, E. M. Wright, W. Sibbett, and K. Dholakia, White light propagation invariant beams, Opt. Express **13**, 6657-6666 (2005).

[2005_21] M. Bock, *Meßsystem für die Bestimmung der spektralen Transferfunktion bei der Strahlformung von Ultrakurzpulslasern*, Diploma Thesis, Technical High School Wildau, 2005 (in German).

[2005_22] P. Saari, M. Menert, and H. Valtna, Photon localization barrier can be overcome, Opt. Commun. **246**, 445-450 (2005).

[2005_23] C. P. Hauri, J. Biegert, U. Keller, B. Schäfer, K. Mann, and G. Marowsky, Validity of wave-front reconstruction and propagation of ultrabroadband pulses measured with a Hartmann-Shack sensor, Opt. Lett. **30**, 1563-1565 (2005).

[2005_24] X. Dong, C. Du, S. Li, C. Wang, and Y. Fu, Control approach for form accuracy of microlenses with continuous relief, Opt. Express **13**, 1353-1360

(2005).

[2005_25] Z. Xu, Y. Kartashov, and L. Torner, Reconfigurable soliton networks optically induced by arrays of nondiffracting Bessel beams, Opt. Express **13**, 1774-1779 (2005).

[2005_26] A. W. Lohmann, H. Knuppertz, and J. Jahns, Fractional Montgomery effect: a self-imaging phenomenon, J. Opt. Soc. Am. A **22**, 1500-1508 (2005).

[2005_27] T. Weitkamp, B. Nöhammer, A. Diaz, C. David, and E. Ziegler, X-ray wavefront analysis and optics characterization with a grating interferometer, Appl. Phys. Lett. **86**, 054101 (2005).

[2005_28] J. Duparré, P. Schreiber, A. Matthes, E. Pshenay-Severin, A. Bräuer, and A. Tünnermann, Microoptical telescope compound eye, Opt. Express **13**, 889-903 (2005).

[2005_29] S. E. Lyshevski, *Nano- and Micro-Electromechanical Systems: Fundamental of Micro- and Nano-Engineering*, CRC Press, Boca Raton, FL, 2005 (2nd edition).

[2005_30] S. B. Blickenstaff, A. M. Sarangan, T. R. Nelson, Jr., K. D. Leedy, and D. L. Agresta, Influence of shadow mask design and deposition methods on nonplanar dielectric material, J. Microlith. Microfab. Microsyst. **4**, 023015-1 (2005).

[2005_31] S. Egger, A. Ilie, Y. Fu, J. Chongsathien, D.-J. Kang, and M. E. Welland, Dynamic shadow mask technique: a universal tool for nanoscience, Nano Lett. **5**, 15-20 (2005).

[2005_32] Y. Zhang, L. Wang, and C. Zheng, Vector propagation of radially polarized Gaussian beams diffracted by an axicon, J. Opt. Soc. Am. A **22**, 2542-2546 (2005).

[2005_33] U. Neumann, R. Grunwald, U. Griebner, G. Steinmeyer, M. Schmidbauer, and W. Seeber, Second harmonic performance of a-axis oriented ZnO nanolayers on sapphire substrates, Appl. Phys. Lett. **87**, 171108 (2005); http://www.virtualjournals.org/vjs/notification.jsp.

[2005_34] S. Reichelt and H. Zappe, Combined Twyman-Green and Mach-Zehnder interferometer for microlens testing, Appl. Opt. **44**, 5786-5792 (2005).

[2005_35] S. Reichelt, A. Bieber, B. Aatz, and H. Zappe, Micro-optics metrology using advanced interferometry, Proc. SPIE Vol. **5856**, 437-446 (2005).

[2005_36] S. Akturk, X. Gu, P. Gabolde, and R. Trebino, The general theory of first-order spatio-temporal distortions of Gaussian pulses and beams, Opt. Express **13**, 8642-8661 (2005).

[2005_37] Prof. H. Weber, Technical University Berlin (private communication).

[2005_38] I. Amat-Roldán, I. G. Cormack, P. Loza-Alvarez, and D. Artigas, Measurement of electric field by interferometric spectral trace observation, Opt. Lett. **30**, 1063-1065 (2005).

[2006_1] J. Liu, B-Y. Gu, J.-S. Ye, and B.-Z. Dong, Applicability of improved

Rayleigh-Sommerfeld method 1 in analyzing the focusing characteristics of cylindrical microlenses, Opt. Commun. **261**, 187-198 (2006).

[2006_2] J. W. Arkwright, Optical-correction technique makes lambda/1000 optics a reality, Laser Focus World, May 2006, 85-89.

[2006_3] P. Ruffieux, T. Scharf, and H. P. Herzig, On the chromatic aberration of microlenses, Opt. Express **14**, 4687-4694 (2006).

[2006_4] O. N. Stavroudis, *The mathematics of geometrical and physical optics*, Wiley-VCH, Weinheim, 2006.

[2006_5] R. Grunwald, U. Neumann, U. Griebner, G. Steinmeyer, G. Stibenz, and M. Bock, Self-reconstruction of pulsed optical X-waves, in: J. A. Kong (Ed.), *Localized Waves, Theory and experiments*, Wiley & Sons, 2006, in press.

[2006_6] R. Grunwald, U. Neumann, A. Rosenfeld, J. Li, and P. R. Herman, Scalable multichannel micromachining with pseudo-nondiffracting vacuum ultraviolet beam arrays generated by thin-film axicons, Optics Letters **31**, 1666-1669 (2006).

[2006_7] R. Grunwald, M. Bock, U. Neumann, G. Stibenz, and G. Steinmeyer, Fringe-free nondiffracting beams for advanced laser diagnostics and processing, CLEO/QELS 2006, Long Beach, May 21-26, 2006, Technical Digest CD-ROM, paper CMX1.

[2006_8] R. Grunwald, M. Bock, G. Steinmeyer, and G. Stibenz, Graxicons for hyperspectral diagnostics of few-cycle laser pulses, CLEO/QELS 2006, Long Beach, May 21-26, 2006, Technical Digest CD-ROM, paper CWJ2.

[2006_9] M. Ferstl, G. Devendra, R. Grunwald, and M. Bock, Lithographically fabricated micro-optical array beam shapers for ultra-short pulse lasers, CLEO/QELS 2006, Long Beach, May 21-26, 2006, Technical Digest CD-ROM, Poster JThC98.

[2006_10] P. Saari, How small a packet of photons can be made?, Laser Physics **16**, 556-561 (2006).

[2006_11] W. C. Soares, D. P. Caetano, and J. M. Hickmann, Hermite-Bessel beams and the geometrical representation of nondiffracting beams with orbital angular momentum, Opt. Express **14**, 4577-4582 (2006).

[2006_12] C. López-Mariscal, J. C. Gutiérrez-Vega, G. Milne, and K. Dholakia, Orbital angular momentum transfer in helical Mathieu beams, Opt. Express **14**, 4182-4187 (2006).

[2006_13] H. Valtna, K. Reivelt, and P. Saari, Modifications of the focused X waves as suitable models of strongly localized waves for realization in the optical domain, J. Opt. A: Pure Appl. Opt. **8**, 118-122 (2006).

[2006_14] Y. V. Kartashov, A. A. Egorov, V. A. Vysloukh, and L. Torner, Shaping soliton properties in Mathieu lattices, Opt. Lett. **31**, 238-240 (2006).

[2006_15] M. Zamboni-Rached, Diffraction-attenuation resistant beams in absorption media, Opt. Express **14**, 1804-1809 (2006).

[2006_16] Microdisplay schaltet 15.300-mal pro Sekunde, Laser Technik Journal, June 2006, No. 3, 28 (2006) (in German).

[2006_17] J. Azaña and S. Gupta, Complete family of periodic Talbot filters for pulse repetition rate multiplication, Opt. Express **14**, 4270-4279 (2006).

[2006_18] Q. Cao and J. Jahns, Apodized multilevel diffractive lenses that produce desired diffraction-limited focal spots, J. Opt. Soc. Am. A **23**, 179-186 (2006).

[2006_19] J. Jahns, A. W. Lohmann, Diffractive-optical processing of temporal signals, part I: basic principles, Chinese Opt. Lett. **4**, 259-261 (2006)

[2006_20] J. Jahns, A. W. Lohmann, Diffractive-optical processing of temporal signals, part II: optical tapped-delay line, Chinese Opt. Lett. **4**, 262-264 (2006)

[2006_21] P. Tabeling, *Introduction to Microfluidics*, Oxford University Press, Oxford, 2006.

[2006_22] M. Bock, Spatio-spectral diagnostics of ultrashort laser pulses by means of statistical moments, Photonik international, March 2007 (in press).

[2006_23] B. J. Chun, C. K. Hwangbo, and J. S. Kim, Optical monitoring of nonquarterwave layers of dielectric multilayer filters using optical admittance, Opt. Express **14**, 2473-2480 (2006).

[2006_24] J. I. Larruquert, Reflectance optimization of inhomogeneous coatings with continuous variation of the complex refractive index, J. Opt. Soc. Am. A **23**, 99-107 (2006).

[2006_25] W. Seeber, U. Neumann, R. Grunwald, J.-P. Mosnier, R. G. O'Haire, and E. McGlynn, High-performance materials based on nonlinear optical composites of ZnO-nanolayers, 8th International Otto Schott Colloquium, July 23-27, 2006, Jena, Germany, Poster, Proceedings P83.

[2006_27] F. Charrière, J. Kühn. T. Colomb, F. Montfort, E. Cuche, Y. Emery, K. Weible, P. Marquet, and C. Depeursinge, Characterization of microlenses by digital holographic microscopy, Appl. Opt. **45**, 829-835 (2006).

[2006_28] A. W. Lohmann, *Optical Information processing*, S. Sinzinger (Ed.), Universitätsverlag Ilmenau, 2006.

[2006_29] J. Yang and H. G. Winful, Generalized eikonal treatment of the Gouy phase shift, Opt. Lett. **31**, 10-106 (2006).

[2006_30] A. Dubietis, G. Valiulis, and A. Varanavicius, Nonlinear localization of light, Lithuanian J. Phys. **46**, 7-18 (2006).

[2006_31] J. Leach, G. M. Gibson, M. J. Padgett, E. Esposito, G. McConnell, A. J. Wright, and J. M. Girkin, Generation of achromatic Bessel beams using a compensated spatial light modulator, Opt. Express **14**, 5581-5587 (2006).

[2006_32] R. Slavik, Y. Park, M. Kulishov, R. Morandotti, and J. Azaña, Ultrafast all-optical differentiators, Opt. Express **14**, 10699-10707 (2006)

Patents

[P1]   R. Grunwald, G. Szczepanski, I. Pinz, and D. Schäfer, Verfahren zur Herstellung optischer Elemente mit örtlich variablem Reflexionsgrad, DD 332 071 5 (1991).

[P2]   R. Grunwald and U. Griebner, Phasengekoppelter Laser mit mehrfach selbst-abbildendem Resonator DE 19502054.5(1996).

[P3]   R. Grunwald, and S. Woggon, Verfahren zur Herstellung anamorphotischer mikrooptischer Arrays und damit ausgestattete Faserlinse, DE 196 13 745 (2004).

[P4]   R. Grunwald, S. Woggon, and U. Neumann, Anordnung zur Strahlformung und räumlich selektiven Detektion mit nicht-sphärischen Mikrolinsen, DE 197 21 257 A1 (2005).

[P5]   R. Grunwald, U. Griebner, E. Büttner, and C. Lukas, Verfahren und Anordnung zur zeitlich und spektral aufgelösten Charakterisierung von ultrakurzen Laserimpulsen, DE 19935630C2 (2003.

[P6]   R. Grunwald, U. Griebner, H.-J. Hartmann, V. Kebbel, T. Elsaesser, and W. Jüptner, Method and arrangement for spatially resolved interferometric characterization of ultrahort laser pulses, US 6,683,691 B2 (2004).

[P7]   R. Grunwald, U. Griebner, H.-J. Hartmann, V. Kebbel, T. Elsaesser, and W. Jüptner, Verfahren und Anordnung zur Charakterisierung ultrakurzer Laserimpulse, DE 199 35 631 (2001).

[P8]   R. Grunwald, U. Griebner, H.-J. Hartmann, V. Kebbel, T. Elsaesser, and W. Jüptner, Verfahren und Anordnung zur orts- und zeitaufgelösten interferometrischen Charakterisierung ultrakurzer Laserimpulse, DE 100 28 756 A1 (2004); EP 01250179.7 (2002).

[P9]   R. Grunwald and H.-J. Kühn, Verfahren und Anordnung zur Erzeugung einer orts- und winkelabhängigen Antireflexionsbeschichtung für mikrooptische Bauelemente, DE 101 13 466 (2002).

[P10]   J.-W. Tomm and R. Grunwald, Verfahren und Anordnung zur Justierung von optischen Bauelementen zwecks Strahlformung von Halbleiterstrahlungsemittern, DE 101 35 101 A1 (2004).

[P11]   A. Fischer and R. Grunwald, Verfahren und Anordnung zur orts- und winkelaufgelösten Reflexionsmessung, DE 10238078 C1 (2003).

[P12]   R. Grunwald, U. Neumann, and V. Kebbel, Verfahren und Anordnung zur ortsaufgelösten Charakterisierung der Krümmung einer Wellenfront, DE 102 43 838 B3 (2004).

[P13]   R. Grunwald, U. Neumann, and T. Elsässer, Verfahren zur Leistungserhöhung von optischen Strahlformern und optischer Strahlformer, DE 10 2004 011 190 A1 (2005).

[P14]   D. Gabor, Improvements in or relating to optical systems composed of lenticules, Patent UK 541 753 (1940).

[P15]   W. B. Hugle, R. Dändliker, H. P. Herzig, Lens array photolithography, UK 9224080.3 (1992) and US 08/114,732 (1993).

[P16]   R. Grunwald, M. Bock, Verfahren und Anordnung zur Charakterisierung der spektralaufgelösten Wellenfront ultrakurzer Laserimpulse (filed 2005).

[P17]   G. J. Swanson and W. B. Veldcamp, High Efficiency, Multi-Level, Diffractive Optical Elements, US Pat. 4895790 (1990).

FIGURE CREDITS

The author gratefully acknowledges the reprint permissions by the copyright holders:

American Physical Society (APS): Fig. 191 [2003_7].

European Optical Society (EOS): Figs. 164-166 [2000_3], Fig. 168 [2001_9], Fig. 145 [2004_15].

Friedrich-Schiller-University, Jena, Dr. B. Kley: Fig. 10 [private communication].

John Wiley & Sons: Fig. 200 [2006_5].

Optical Society of America (OSA): Figs. 87-96 [1999_2], Fig. 167 [2000_4], Figs. 181-183 [2003_1], Figs. 153, 154, 157 [2004_1], Fig. 148 [2004_5], Figs. 151, 180, 185 [2005_11], Fig. 193 [2005_12], Fig. 192 [2006_5], Figs. 160, 162 [2006_6], Fig. 146 [2006_7], Figs. 75, 188-190 [2006_8].

Springer: Figs. 46, 64 [1997_5]

The International Society for Optical Engineering (SPIE): Figs. 52, 53, 55, 56, 110 [2001_1], Figs. 97-103 [2001_11], Figs. 108, 109, 111-115 [2004_2], Figs. 139, 141 [2004_11], Figs. 149, 184 [2005_8].

Wiley-VCH: Figs. 73,135 [2004_8], Figs. 32,74 [2005_6].

GLOSSARY

Aberrations

Aberrations describe deviations from the ideal imaging performance of an optical system for a given input wave field. They are classified with respect to their geometrical and physical origin (spherical aberrations, chromatic aberrations, polarization aberrations etc.). Compared to macroscopic lenses of the same radius of curvature, microlenses have smaller spherical aberrations because of their much narrower angular spectra. Depending on the basic optical model applied (geometrical optics, wave optics), ray aberrations or wave aberrations are distinguished. See also: ▶ Zernike polynomials and ▶ Seidel coefficients.

Adaptive mirror

Flexible mirror of steerable curvature. Adaptive mirrors consist of continuous or segmented membranes which are actuated by piezo-translators, electromagnetic fields, acoustically or by heating.

Advanced Shack-Hartmann wavefront sensor

▶ Shack-Hartmann sensor with ▶ axicons.

AFM

▶ Atomic force microscope

Airy disc

Central maximum of the 2D far-field intensity distribution of the diffraction pattern of a hard circular aperture.

Ambiguity function

Characteristic function of an optical system [1972_5, 2005_3, 2000_2] which can be used to describe the 3D ▶ optical transfer function and serves as a merit function for the design of optical systems with large depth of focus [2005_4].

Amplitude apodization

▶ Apodization by adapted amplitude masks based on space-variant absorption or reflection.

Anamorphotic

Imaging systems with differing divergence in two spatial directions (i.e. without rotational symmetry), in particular systems of cylindrical lenses which commonly are used to collimate elliptical, astigmatic beams (see also: ▶ astigmatism) of laser diodes. From Greek "anamorphosis" (reshaping).

Angular dispersion

This type of dispersion is characterized by a change in the propagation direction (angular dispersion) or spatially variable ▶ pulse fronts (pulse front tilt, pulse front distortion) by different travel time within media of different thickness or propagation in different angles. See also: ▶ dispersion, ▶ geometrical dispersion and ▶ chromatic dispersion.

Apodization

Partial suppression of side-lobes of diffraction patterns by modified pupil functions (see also: ▶ Toraldo concept, ▶ phase apodization, ▶ amplitude apodization).

Array generator

Microoptical systems designed for the transformation of beams into ▶ beam arrays.

Astigmatism

Particular class of ▶ aberrations. Often, it is indicated by the appearance of different focal points for rays propagating in perpendicular planes. Different kinds of astigmatism are known depending on the symmetry of the optical system. Third-order astigmatism ("monochromatic astigmatism") is related to radially symmetric systems, whereas other kinds are related to non-radially symmetric systems.

Atomic force microscope

Highly resolving microscope for 3D surface inspection consisting of a nanotip at the end of a cantilever which is deflected by Van der Waals force. The deflection is measured optically or piezoelectrically. AFM can be operated in contact mode (constant force by keeping the deflection constant), non-contact mode (cantilever oscillates near resonance frequency), and dynamic contact mode (cantilever periodically in contact). In contrast to scanning electron microscopes, no vacuum system is necessary, but fields of view and depth of field are smaller.

Autocorrelation

The autocorrelation function represents the ▶ cross-correlation function of a signal with itself. It is a Hermitian operator. First order autocorrelation delivers a symmetric autocorrelation function. Information about signal asymmetry can be derived by higher order and ▶ triple autocorrelation. See also: ▶ collinear autocorrelation, ▶ cross-correlation, ▶ intensity autocorrelation, ▶ interferometric autocorrelation, and ▶ non-collinear autocorrelation, and ▶ triple autocorrelation.

Autocovariance function

The autocovariance function is the covariance of a distribution (in particular a phase profile of a surface in interferometry) with itself. Mathematically it is closely related to the ▶ autocorrelation function and can be used to evaluate the quality of periodical structures (arrays) [1999_5]. See also: ▶ correlation length and ▶ surface spatial wavelength ([1993_6], pp. 44-47).

Axicon

Optical element like a conical lens, a conical mirror or a ▶ Computer-generated hologram transforming a beam into an extended focal zone. Originally, axicons were proposed by McLeod for slide projectors to ensure a sharp image independently on the axial position [1954_1] ("constant focus along the axis"); see also: ▶ Gaussian axicon and ▶ microaxicon.

Axicon-based Shack-Hartmann sensor

▶ Shack-Hartmann sensor with ▶ axicons.

Backside coated chirped mirror

To overcome the problem of impedance matching between a ▶ chirped mirror and an AR coating, the concept of backside-coated chirped mirror (BASIC) was introduced. Here, the mirror coating is placed on the back instead of front of a thin wedged glass substrate [2000_6].

Bandwidth-limited

See: ▶ wavepacket, ▶ time-bandwidth product and ▶ space-bandwidth product.

BASIC

▶ Backside coated chirped mirror.

Beam array

One- or two-dimensional arrangement (matrix) of beams generated by arrays of microoptical elements like microlenses or microaxicons.

Beam parameter product

Measure of the spatio-angular quality of a Gaussian beam given by the product of beam waist radius and total divergence angle: $BP = w_0\, \tau = 2\lambda/\pi$.

Beam propagation factor

The beam propagation factor K is the inverse of the ▶ beam quality factor M^2 [2000_39].

Beam quality factor

The beam quality factor M^2 is a standard measure of the beam quality and the ▶ inverse of the beam propagation factor K [2000_39]. It describes the deviation from an ideal Gaussian beam. If the propagating field is generated in a laser resonator, it is typically related to the number of oscillating transversal modes. The determination of its value requires to measure the waist diameter $d_0 = 2w_0$ and divergence angle θ with respect to the corresponding second moments of the energy distribution: $M^2 = (\pi/\lambda)(w_0\theta/2)$.

Bessel beam

Particular type of ▶ nondiffracting beam (or ▶ diffraction-free beam) obtained as a propagation invariant solution of the ▶ Helmholtz wave equation for conical waves. The amplitude distribution of a Bessel beam follows a zero-order Bessel function and shows characteristic concentric fringes. An ideal (theoretical) Bessel beam is infinitely extended and carries infinite energy. All of its fringes have equal energy. Real-world experiments approximate Bessel beams by ▶ Bessel-like beams (or pseudo-Bessel beams) of finite energy and radial extension (see also: ▶ undistorted progressive waves, ▶ X-wave, ▶ X-pulse, ▶ Mathieu beam, ▶ needle beam).

Bessel-Gauss beam

Pseudo-nondiffracting beam generated by illuminating an ▶ axicon with a beam of Gaussian intensity distribution in transversal direction.

Bessel-Gauss pulse

Pulsed (polychromatic) ▶ Bessel-Gauss beam.

Bessel-like beam

Physically realizable ▶ pseudo-nondiffracting beam similar to a ▶ Bessel beam but with finite energy (see also: ▶ pseudo-Bessel beam). In contrast to ideal Bessel beams, the spatial frequency of its fringes can vary with the propagation distance [1991_5].

Bessel pulse

Pulsed, polychromatic ▶ Bessel beam. Ultrashort pulsed Bessel beams of sufficiently broad spectra (e.g. for Ti:sapphire laser at pulse durations < 10 fs) lead to pulsed ▶ X-waves (▶ X-pulses) because of ▶ spectral interference.

Bessel-X-wave

▶ X-wave of ▶Bessel-like beam structure generated by spectral interference of broadband conical radiation.

Binary optics

Flat diffractive optical structures were referred to as binary optics by Swanson and Veldkamp [P17].

Blazed structures

The term "blazing" is used on different levels. Blazed gratings are reflection gratings with asymmetric grooves. A more specific meaning of "blazing" is the thickness reduction by sampling the phase function of optical components at multiples of 2π ([1999_1], p. 130). See also: ▶ Micro-Fresnel lens.

Bragg mirror

Highly reflecting mirror based on Bragg reflection at a periodic index structure (Bragg reflector). A stack of layers of transparent optical materials of alternating refractive indices is designed to have optical thicknesses of a quarter wavelength for each layer (quarter-wave-mirror). The reflectivity of a Bragg mirror results from the constructive multiple interference of the partial waves produced by Fresnel reflection at all interfaces of the layers. The reflectivity is determined by the number and index difference of the layer pairs. The ▶ group delay dispersion is minimal only at the design wavelength. Therefore, ▶ chirped mirrors have to be used for ultrashort pulses of large bandwidth.

Carrier-envelope offset

▶ Carrier envelope phase (CEP) offset: Offset of the phase of time-dependent field amplitude of an ultrashort pulse with respect to the phase of the envelope function of the wavepacket.

Carrier envelope phase (CEP)

Phase of the temporal envelope of the carrier wave of an ultrashort ▶ wavepacket. See also: ▶ Carrier-envelope offset.

CEO

▶ Carrier-envelope offset.

CEP
▶ Carrier envelope phase

CGH
▶ Computer-generated hologram.

Chirp
Frequency variation of a signal as a function of time (▶ temporal chirp) or space (▶ spatial chirp) or space and time (▶ spatio-temporal chirp); see also ▶ space-time-coupling.

Chirped mirrors
Chirped mirrors are ▶ Bragg mirrors of variable layer thickness to minimize the ▶ group delay dispersion of ultrashort pulses [1994_27]. The design is adapted to the spectral phase (reflector of different depth for different spectral contributions; ▶ double chirped mirrors, ▶ backside coated chirped mirror).

Chromatic aberration
Aberration caused by ▶ chromatic dispersion. In particular in the case of extremely broad spectral bandwidth, the design of optical systems can hardly compensate for imaging errors over the complete spectral range.

Chromatic dispersion
Dispersion caused by different optical pathways for different parts of the spectrum in a refractive medium. Chromatic dispersion can lead to a temporal broadening of ultrashort pulses. Other dispersion effects are ▶ angular dispersion or ▶ geometrical dispersion. Nonlinear index changes at high intensities lead to nonlinear dispersion effects.

Coherent control
Control of the temporal dynamics of chemical reactions by pump-probe-type experiments. The coherence is destructed by specific interactions (▶ decoherence). Quantum interference between different excitation pathways is used to eliminate unwanted non-radiative channels. Thus, reaction-path-selective chemistry can be realized [2001_30].

Collinear autocorrelation
▶ Autocorrelation setup where the beam and its replica are collimated in one axis. See also: ▶ noncollinear autocorrelation..

Columnar thin film
Thin optical films of columnar local structure.

Computer-generated holograms
Computer generated holograms (CGH) are holographic beam shaping elements the structure of which was designed by a numerical optimization of the diffraction pattern in a target plane. CGHs enable a tailoring of wavefronts without the need for classical photographic recording. They can be realized as fixed phase and/or amplitude patterns transferred into binary optical hardware as well as dynamic patterns from programmable high-resolution ▶ spatial light modulators.

Confocal parameter

Parameter describing the depth of focus by twice the ▶ Rayleigh length. For an ideal Gaussian beam, the confocal parameter is $b = \pi w_0^2/\lambda$. In terms of the ▶ beam parameter product BP we can write $b = BP/\theta^2$.

Correlation length

Characteristic length in the spatial ▶ autocovariance function (e.g. of a surface profile) which corresponds to a reduction of this function by 1/e. See also: ▶ surface spatial wavelength ([1993_6], pp. 44-47).

Cross-correlation

Mutual correlation of two data sets; see also: ▶ autocorrelation.

Cross-talk

Energy transfer or mode coupling between neighboring channels in multichannel transmission systems, in particular at high channel density. In free space systems with ▶ microlens arrays, cross-talk mainly arises from diffraction, spherical and chromatic aberrations, scattering and the overlap of diverging subbeams; see also: ▶ parallel processing.

CTF

▶ Columnar thin films.

Cylindrical axicon

In analogy to "cylindrical lens", an element with a generalized cylindrical surface shape which enables to generate an extended zone of quasi-nondiffracting beam propagation. A special case is the Fresnel's biprism. See also: ▶ axicon.

Dammann grating

Two-level binary phase grating; see also: ▶ kinoform.

Decoherence

Non-radiative processes like vibrational relaxation can destruct the coherence information of a system. With a ▶ coherent control by spectral or temporal selection, a suppression of such pathways and an improvement of the state-selectivity are possible.

Dichroic

See: ▶ dichroism.

Dichroism

The term "dichroism" is applied to different wavelength selective effects. The first one describes the splitting of a beam into distinct beams of different wavelengths (e.g. by a multilayer system), whereas the second one is related to a selective absorption (e.g. of circularly polarized light) resulting in a particularly polarized beam.

Diffraction-free beams

Propagation-invariant solutions of Helmholtz wave equation (▶ Bessel beams, ▶ Mathieu beams etc.). Experimentally, Bessel beams can be approximated by the Fourier transform of circular slits or generating beams of conical angular distribution behind refractive, reflective or diffractive ▶ axicons. A propagation invariance (in real-world experiments appearing over a limited distance) results from the successive interference of different partial waves. Therefore, the terms "non-spreading beams" or ▶ "undistorted progressive waves" are more adequate to describe the phenomenon.

Diffraction-limited optical systems

Idealized microoptical systems the spatio-temporal resolution of which is only limited by the diffraction (see also: ▶ wavepacket, ▶ transform-limited pulse, ▶ time-bandwidth product and ▶ space-bandwidth product).

Diffractive optical elements (DOE)

Beam shaping elements based on diffraction with feature sizes of substructures comparable to or smaller than the wavelength. DOE can be implemented as continuous relief or binary structures; see also: ▶ Dammann grating and ▶ kinoform.

Diffuse reflectance

Part of the light reflected from an object surface in directions out of the one given by the law of reflection of geometrical optics. Diffuse and ▶ specular reflectance can not completely be separated because of their overlapping spatial characteristics. In contrast to the specular reflectance, angular-dependent changes in the spectral distribution of broadband signals can appear.

Digital holography

Holography based on the digital reconstruction of a phase and amplitude object in a computer. Instead of illuminating a hologram by a reference wave, a detected interference pattern is electronically processed with digital reference data.

Digital light processor^{TM}

Digital light processor^{TM} (DLP^{TM}) technology is a part of Texas instruments and stands for the second generation of highly integrated digital micromirror devices (▶ DMD^{TM}). Arrays of flipping micromirrors are fabricated on a single, semiconductor-based all-digital display chip; see also: ▶ spatial light modulator.

Digital micromirror device^{TM}

Name of the first generation of digital micromirror devices (trademark of Texas Instruments); see also: ▶ MEMS, ▶ digital light processor, ▶ spatial light modulator.

Dispersion

Dispersion is a measure for the separation of spectral and/or angular parts of a light beam by the interaction with a dispersive medium (▶ chromatic dispersion, ▶ nonlinear dispersion ▶ angular dispersion).

DLP^{TM}

▶ Digital light processor^{TM}.

DMD^{TM}

▶ Digital micromirror device^{TM}.

DMM

Digital ▶ micromirrors; see also: digital light processor.

DOE

▶ Diffractive optical elements.

Double chirped mirrors
▶ Chirped mirrors with two envelope functions of the layer thicknesses. One of these envelopes corresponds to the usual ▶ chirped mirror design. See also: ▶ Backside coated chirped mirror.

Effective index method
For the simulation of light propagation through multilayer structures, calculation methods based on a matrix representation with effective optical constants were developed [2003_12].

Elliptical nondiffracting beams
Generalization of ▶nondiffracting beams to non-rotationally-symmetric beams [2002_21], see also: ▶ Mathieu beams.

EUV
▶ Extreme ultraviolet.

Extreme ultraviolet
The spectral range between VUV and X-rays (wavelengths of about 1-50 nm); see also: ▶ vacuum ultraviolet.

Fabry-Pérot interferometer
This special type of interferometer or resonator, also called etalon, consists of two parallel, partially reflecting mirrors which can be the surfaces of a transparent plate or separated elements. Optical ▶ multilayer systems are multiple thin etalons on a substrate.

Fill factor
Measure of the geometrical filling of an array, typically a microlens array, by the ratio of filled area and total area. In some cases, it can be used to compare the quality of fabrication technologies. Because of effects like aberrations, diffraction or parasitic waveguiding (causing cross talk and losses), however, the fill factor is not a reliable measure of the efficiency of microoptical systems.

Focus mode waves
Particular type of localized solutions of the three-dimensional wave equation. Their characteristics can be adjusted by a free parameter between being a plane wave and a spatially localized pulse [1983_3].

Fourier limit
See: ▶ wavepacket.

FP
▶ Fabry-Pérot interferometer.

Frequency resolved optical gating (FROG)
Well-established method for the characterization of temporal shape and ▶ spectral phase of ultrashort pulses based on a measurement of the background-free ▶ autocorrelation and the spectrum [2000_9]. Here, an unknown probe pulse samples an unknown gate function by a nonlinear interaction. Different types of FROG are related to different nonlinear processes (▶ second harmonic generation, ▶ third harmonic generation, self-diffraction, non-linear polarization). See also: ▶ GRENOUILLE.

FROG

▶ Frequency resolved optical gating.

FWHM

Full width at half maximum.

Gabor superlens

Microoptical system formed by a pair of microlens arrays which slightly differ in their pitches. The local pairs of microlenses act as beam steering elements generating converging subbeams so that the whole system works as a facetted lens [P14, 1999_21].

Gaussian axicon

Refractive ▶ axicon of Gaussian-shaped, convex phase profile. Its reflective counterpart is a concave axicon of inverse Gaussian phase profile.

GDD

▶ Group delay dispersion.

Geometrical dispersion

Dispersion effect of particular relevance for ultrashort pulses which is caused by different travel times in thick media or at different propagation angles. Geometrical dispersion leads to pulse front tilts or distortions. See also: ▶ dispersion, ▶ chromatic dispersion and ▶ angular dispersion.

Gerchberg-Saxton algorithm

Mathematical algorithm for the phase retrieval from a pair of intensity distributions of a propagating light beam related to a function governing this propagation (e.g. Fourier transform).

Gires-Tournois interferometer

Interferometer consisting of a highly reflecting (HR) mirror and a partially reflecting mirror typically used to control the ▶ group delay dispersion (GDD) [1964_1; 2003_26, p. 395]. Devices with an airgap enable for a tuning of the GDD. This setup can be regarded as a special case of a folded ▶ Fabry-Pérot interferometer. If the GTI consists of a dielectric layer on top of a HR mirror, however, one has to take into account an additional phase step at the interface of HR mirror and layer caused by the spectral absorption of the HR mirror. To avoid parasitic GTI effects, hybrid refractive-refractive elements have to be operated at significantly large angles (not in normal incidence) or with a broadband AR-coating. A better control of dispersion can be achieved with ▶ chirped mirrors.

Goos-Haenchen-Effekt

Longitudinal displacement of a wave at total internal reflection. The effect can be positive (as in most practical cases) or negative depending on the conditions.

Gouy phase

For conical beams like Gaussian or Bessel beams and focused parts of plane waves, the phase along the axis follows a nonlinear distribution compared to the linear case of a plane wave. This phase difference is called Gouy phase shift [1890_1, 1890_2,

1980_3]. The Gouy phase shift can be understood as a geometrical quantum effect (transverse spatial confinement of photons) [1996_29, 2001_36].

Graxicon

Hybrid optical element combining the diffractive properties of a grating with the refractive properties of an ▶ axicon [2006_8].

Graded reflectance micro-mirror array

Array of miniaturized ▶ Graded reflectance mirrors with or without envelope functions of phase and/or reflectivity.

Graded reflectance mirror

Mirror of spatially variable reflectivity coefficient typically consisting of a multilayer containing one ore more layers of space-variant thickness (see also: ▶ Graded reflectance micro-mirror array).

Gradient index lenses

Lenses with a spatially nonuniform refractive index. In most cases, a glass rod with radial index profile is used to obtain imaging properties without any surface curvature.

Graded spectral filter

Spectral filter of space-variant spectral characteristics (e.g. by a spatially variable ▶ multilayer system); see also: ▶ graded reflectance mirror, ▶ graded reflectance micro-mirror array.

Grenouille

"Grating-eliminated no-nonsense observation of ultrafast incident laser light E-fields". Modified version of ▶ FROG device where the dispersion of a prism is used to obtain the spectral resolution [2001_38].

GRIN lenses

See: ▶ gradient index lenses.

Grism

Combination of a (surface or volume) grating with a prism. Light of a certain design wavelength is kept undeviated whereas the other parts of the spectrum are refracted. Grisms are applied to stretcher-compressor stages of ultrashort-pulse laser systems and compact astronomical spectrometers. If an object is imaged onto a camera, a grism in the collimated beam leads to a spectrally dispersed image; see also: ▶ graxicon.

GRM

▶ Graded reflectance mirror.

GRM

▶ Graded reflectance micro-mirror array.

Group delay

The group delay is the negative derivative of the phase shift with respect to the angular frequency ($-d\varphi/d\omega$) in units of time. In many systems the optimum performance is obtained if the group delay does not depend on the frequency over a spectral interval of interest (e.g. transmission systems for ultrashort pulses). See also: ▶ group delay dispersion.

Group delay dispersion

Spectral dependence of the ▶ group delay (also called second-order dispersion). It is given by the derivative of the group delay with respect to the angular frequency and thus equivalent to the second derivative of the phase with respect to the angular frequency ($d^2\varphi/d\omega^2$) in units of time^2 (e.g., [fs^2]).

Group velocity

Propagation velocity of the envelope of a wave. It is the inverse of the derivative of the wave number with respect to the angular frequency $1/(dk/d\omega)$. See also: ▶ superluminal.

Group velocity dispersion

Derivative of the ▶ group velocity with respect to the angular frequency.

GTI

   ▶ Gires-Tournois interferometer.

GVD

   ▶ Group velocity dispersion.

Hartmann-sensor

Wavefront sensor based on partitioning a beam into separated sub-beams by an array of sub-apertures (holes). From analyzing the displacement of the centers of gravity of the sub-beams, information on the local tilt can be obtained (see also: ▶ Wavefront sensor and ▶ Shack-Hartmann sensor).

Hermite-Gaussian beam

Orthogonal set of solutions of the Helmholtz wave equation corresponding to the well-known TEM$_{mn}$-laser modes.

HHG

   ▶ Higher harmonics generation.

Higher harmonics generation

At high intensities, higher order harmonics of the laser wavelengths can be generated via multiphoton processes. The simultaneous excitation of atomic or molecular oscillators by multiple photons enables them to emit photons of higher energy.

Holographic optical element (HOE)

Holographic optical elements are generated by a holographic recording process (fixing the complex wavefield properties in a material by simultaneous exposure with an object wave and a reference wave).

Hopkins ratio

Ratio of the ▶ modulation transfer functions (MTF) of an imaging system at a given spatial frequency with and without aberrations [1998_2, p. 182].

Hybrid microoptical elements

Microoptical elements combining two or more working principles (e.g. refraction plus diffraction, reflection plus refraction or reflection plus diffraction).

Hypotrochoidal curve

Here: orbit of a substrate point in a deposition system with planetary rotation.

Intensity autocorrelation

▶ Autocorrelation exploiting the transversal envelope function of a noncollinear autocorrelation pattern (in most cases without resolving the interference fringes); see also: ▶ interferometric autocorrelation.

Interferometric autocorrelation

▶ Collinear autocorrelation method with an interferometric setup resolving interference fringes in propagation direction; see also: ▶ intensity autocorrelation.

Kerr effect

The nonlinear optical Kerr effect or quadratic electrooptic effect describes the dependence of the refractive index on the square of the electrical field (in contrast to the Pockels effect where it depends linearly). The measure of the Kerr effect is the Kerr nonlinearity or nonlinear refractive index (n_2) which changes at high intensities I. The resulting total refractive index is $n = n_0 + n_2 I$ where n_0 is the refractive index at low intensity.

Kinoform

Continuous or multilevel stairstep surface-relief phase plates or phase holograms for the shaping of arbitrary intensity distributions in a focal plane. For the computer generated design of kinoforms, iterative Fourier transform algorithms are used.

Kramers-Kronig relation

The Kramers-Kronig relation expresses the connection between real and imaginary part of a function. In optics it is typically applied to relate refractive index and absorption to each other [2005_14].

Laguerre-Gauss beam

Laguerre-Gaussian beams are a special solution of Maxwell equations with an optical vortex in the center and correspond to Laguerre-Gaussian polynomials.

Laser chemical vapor deposition

Deposition of materials by laser-induced chemical processes. In particular, nonthermal and selective pathways for ionic or neutral products can be opened by photoreactions or photo-dissociation. Via a short-pulse multi-photon excitation of molecules, highly reactive fragments can be produced which enable for a fabrication of thin films.

Laser physical vapor deposition

Deposition of materials by a laser-induced physical (thermal or non-thermal) process (ablation, coulomb explosion, sputtering) on solid state targets; see also: ▶ pulsed laser deposition.

LCVD

▶ Laser chemical vapor deposition.

LIGA

Acronym for the German terms "Lithographie" (X-ray lithography), "Galvanik" (electrodeposition) and "Abformung" (molding) which stand for a multistep fabrication technique for microoptical structures [1995_23].

LPVD

▶ Laser physical vapor deposition.

Luneburg lens

Luneburg lenses are spherical lenses which transform the light of any point source on the surface of a sphere into a plane wave on the opposite side and vice versa.

M^2

▶ Beam quality factor

Maréchal criterion

Optical tolerance criterion related to 1/14 of the wavelength.

Mathieu beam

Particular type of ▶ elliptical nondiffracting beams similar to a ▶ Bessel beam but not rotationally symmetric [2000_20].

Matrix processor

Optical multichannel processor based on arrays of ▶ microlenses or ▶ microaxicons.

MEFISTO

Measurement of electric field by interferometric spectral trace observation (MEFISTO) [2005_38].

MEMS

Micro-electro-mechanical systems. A well-known example are addressable high-definition micromirror arrays based on a large number of separately deformable or tiltable miniaturized reflectors (e.g. ▶ DMD^{TM}, ▶ DLP^{TM}).

MFL

▶ Micro-Fresnel lens.

Microaxicon

Miniaturized ▶ axicon.

Micro-Fresnel lens

Miniaturized Fresnel zone lenses with blazed phase structures (see: ▶ blazed structures). Compared to macroscopic refractive zone lenses, the contribution of diffraction is more significant (in particular, by the outer zones of smallest width). Therefore, micro-Fresnel lenses are often assigned to ▶ diffractive optical elements.

Micro-graxicon

Miniaturized ▶ graxicon.

Microlens

Miniaturized lens (typical diameters < 1 mm, see also: ▶ microlens array).

Microlens array

Array of ▶ microlenses mostly in orthogonal, hexagonal or linear arrangements.

Micromirror

Microoptical or ▶ MOEMS component with reflection as dominating working principle. Addressable micromirror arrays are based on a mechanical movement of reflecting parts, e.g. by shifting, bulging or periodical tilting (▶ DMD^{TM}, ▶ DLP^{TM}) or programming index or dispersion profiles (e.g. by controlling the carrier density in semiconductor devices). Complex phase-amplitude profiles can be realized with multilayer structures (see: ▶ multilayer microoptics).

Microprism

Miniaturized optical prisms.

Modulation transfer function

The modulation transfer function (MTF) describes the contrast factor of a transferred signal field (image) as a function of the spatial frequency. In a generalized description, the MTF is the amplitude-related part of a complex ▶ Optical Transfer Function (OTF) which further contains a ▶ Phase Transfer Function (PTF).

MOEMS

Micro-opto-electro-mechanical systems combining the properties of miniaturized optical, electrical and mechanical components (▶ MEMS).

Montgomery effect

This effect represents a *generalization* of the ▶ Talbot effect. The Talbot self-imaging works in paraxial case with lateral periodicity. Montgomery defined the conditions for an angular spectrum to fulfill the phase matching condition of constructive interference also for nonparaxial beams. He searched for objects (referred to as *Montgomery objects*) which generate periodical axial field distributions. Montgomery could show that there exist discrete spatial frequencies where self-imaging occurs. In 2D frequency space, the allowed spatial frequencies of the object form concentric rings [1967_1, 1968_1, 2003_23].

MTF

▶ Modulation transfer function.

Multichannel processing

See ▶ parallel processing.

Multilayer

System of thin, stacked dielectric layers acting as coupled ▶ Fabry-Pérot interferometers with common resonance modes. Refractive indices and thicknesses are designed to match a resulting target amplitude and phase distribution in transmission or reflection mode. In modelling thin-film multilayer systems, multiple reflections and material composition lead to effective values for refractive index and absorption coefficient; see also: ▶ effective index method.

Multilayer microoptics

Microoptical components consisting of structured ▶ multilayer systems. Additional degrees of freedom are obtained by exploiting the spectral phase properties of such elements; see also: ▶ graded reflectance micro-mirror array, ▶ graded reflectance mirror and ▶ graded spectral filter.

Nanolayer microoptics

Thin-film microoptics based on structured layers with structure depths below 1 µm.

Nanooptics

Optical components and systems with feature sizes or apertures in nanometer range.

Needle beams

Beams of ▶ pseudo-nondiffracting propagation characteristics with only a single intensity maximum instead of fringes ("single-maximum nondiffracting beams"). Such beams can be generated by self-apodized ▶ truncation of ▶ Bessel-like beams or generating such beams at sufficiently small conical angles. The term should not be confused with "Nadelstrahlen" used by Einstein to describe directionality and momentum of photons.

NOEMS

Nano-opto-electro-mechanical systems combining optical functionality with nano-mechanical functionality (▶ MOEMS, ▶ MEMS).

Non-collinear autocorrelation

▶ Autocorrelation setup where the beam and its replica are crossing under a non-zero angle; see also: ▶ collinear autocorrelation.

Nondiffracting beam

By constructive interference of coherent conical beams, axially extended zones of quasi- nondiffracting propagation behavior can be generated. In the literature, there exist many different terms for this kind of beams ("undistorted progressive waves", "diffraction-free beams", "diffraction-less beams", "nonspreading beams" etc.). Many of these terms are misleading because nondiffracting beams are not free of diffraction. see: ▶ Bessel beam, ▶ Bessel-like beam, ▶ Mathieu beam.

Nondiffracting beam array

Matrix of nondiffracting or ▶ diffraction-free beams.

Nondiffracting image

Transport of two-dimensional image information by an array of nondiffracting beams with encoded image data.

Nonlinear refractive index

See: ▶ Kerr effect.

Nonspreading beams

See: ▶ diffraction-free beams, ▶ Bessel beams, ▶ Bessel-like beams, ▶ Mathieu beams.

Octave-spanning spectrum

Spectrum with a bandwidth as large as an optical octave (corresponding to a factor two between minimum and maximum wavelength). It has to be noticed that this definition works well with the ▶ FWHM of smooth spectra. In realistic cases of complex spectra, higher order statistical moments have to be used as a criterion to compare spectral bandwidths.

Optical transfer function (OTF)

Generalized complex transfer function comprising a part related to the amplitude (▶ modulation transfer function) and a phase part (▶ phase transfer function). For ultrashort wavepackets and broadband sources, additional transfer functions like the ▶ spectral transfer function and temporal transfer function are relevant.

Optical tweezers

The forces that localized light (in laser foci or pseudo-nondiffracting beams) acts on matter can be used to trap and to manipulate dielectric particles like cells (e.g. by gripping and moving them like with tweezers). See also: optical ▶ trap.

Orbital angular momentum

Two parts contribute to the angular momentum of an electromagnetic field: (a) the spin caused by circular polarization and (b) an orbital momentum caused by the azimuthal phase of the complex electric field [2003_27]. Beams with helical wavefronts have an orbital momentum in the direction of propagation.

OTF

▶ Optical transfer function.

Parallel processing

Pipelining of optical information or simultaneous processing of materials via multiple channels generated by arrayed microoptical components (e.g. ▶ microlenses, ▶ microaxicons, fibers, waveguides). The channels can be well separated or coupled. Static architectures or addressable configurations can be designed. Applications like parallel data processing or cost-effective high-speed microstructuring of workpieces are enabled by the concurrency of light guiding and/or light-matter interaction zones; see also: ▶ array generator, ▶ beam array, ▶ cross talk.

Phase apodization

▶ Apodization by adapted phase profiles; see also: ▶ amplitude apodization.

Phase shifting interferometer

By continuously varying the phase (e.g. by translating the interferometer arm or tuning a wavelength), the depth resolution of an interferometer can be drastically improved.

Phase transfer function

The phase transfer function describes the transfer of a phase information of an optical wave in an imaging optical system; see also: ▶ modulation transfer function and ▶ optical transfer function.

Planar microoptics

Approaches for a more compact system design by integrating microoptical components in stacked functional planes or double-sided structured substrates. The architecture is challenging with respect to aberration control because of the necessary off-axis operation for multiple reflections (zigzag path).

PLD

▶ Pulsed laser deposition.

Point spread function

The point spread function (PSF) represents the optical image of a point source. Mathematically it is described by the Fourier transform of the ▶ pupil function.

Poynting vector

The Poynting vector S is the cross product of electric and magnetic field components and describes the energy flux of the electromagnetic field.

PSI

▶ Phase shifting interferometer.

Pseudo-Bessel beam

Physically realizable approximation of a ▶ Bessel beam.

Pseudo-nondiffracting beam

Real-world approximation of a ▶ nondiffracting beam; see also: ▶ pseudo-Bessel beam.

PTF

▶ Phase transfer function.

Pulsed Bessel beam

See: ▶ Bessel pulse.

Pulsed laser deposition

Deposition of layers by evaporating a material with a pulsed laser; see also: ▶ laser physical vapor deposition.

Pulse front

Contrary to a ▶ wavefront, a pulse front comprises the points of equal phase of an ultrashort-pulse wavepacket for which the corresponding ▶ Poynting vectors are parallel to each other. A pulse front tilt results from local time delays in axial direction without a change of the propagation direction (e.g. by spectral dispersion in absence of refraction effects).

Pump-probe technique

Technique to probe dynamical processes in matter like excited state relaxation or specific energy transfer by pumping and probing with ultrafast laser pulses of variable delay.

Pupil function

Complex function describing the phase and amplitude properties within the pupil plane of an optical system.

Quarter-wave mirror

See: ▶ Bragg mirror.

Rayleigh criterion

Also Rayleigh limit. Optical tolerance criterion related to 1/4 of the wavelength. Historically Lord Rayleigh showed that a telescope can not be distinguished from a theoretically perfect optical system if it is designed within this tolerance. The criterion is further used to define the limits for the resolution of spectral lines or a wavefront planarity.

Rayleigh length

The Rayleigh length (also: Rayleigh range) describes the depth of focus of a Gaussian beam as the distance between the beam waist and a plane where the area of the cross section of the beam has doubled. This distance corresponds to an increase of the beam radius (i.e. the radial coordinate where the intensity distribution function in a plane perpendicular to the optical axis is reduced to $1/e^2$ of its maximum value) by a factor $\sqrt{2}$. For a Gaussian beam, it is identical to half of the ▶ confocal parameter.

Rayleigh Sommerfeld diffraction theory

Particular scalar theory of diffraction. The Rayleigh-Sommerfeld formula describes the propagation of a wave from one plane to another [1998_2].

RIE

Reactive ion etching.

RMS

The root mean square (RMS) represents the standard deviation which is the square root of the second statistical moment or the square root of the variance [2006_22].

SBP

▶ Space-bandwidth product.

Sculptured thin films

Thin optical films consisting of particular types of columnar substructures. Typically, such structures show optical activity [1997_34, 2005_1].

Second harmonic

See: ▶ second harmonic generation.

Second harmonic generation

Nonlinear optical process resulting in a frequency conversion. The interaction of light with a material enables to combine the energy of pairs of photons and thus to generate photons of twice the frequency of the fundamental beam (frequency doubling). The efficiency of SHG depends on the square of the intensity of the fundamental beam.

Second-order dispersion

See: ▶ group delay dispersion.

Seidel coefficients

Coefficients describing the first five of the third-order wave aberrations calculated by Seidel in 1856. See also: ▶ Zernike polynomials.

Self-imaging

If periodical phase and/or amplitude objects are illuminated with coherent light, the spatial periodicity of the wavefield is replicated at distinct planes by constructive interference. Special cases are ▶ Talbot effect and ▶ Montgomery effect. Self-imaging also appears in temporal [2004_22] and spectral domain.

SESAM

Semiconductor saturable absorber mirrors. Because of the fast saturation of absorptive transitions in semiconductor multi-quantum wells, SESAMs are used for passive mode-locking in ultrashort-pulse laser systems.

SFG

▶ Sum frequency generation.

Shack-Hartmann sensor

Wavefront sensor based on the displacement of focal spots of an array of microlenses with respect to a reference wave (see also: ▶ Hartmann sensor; ▶ Wavefront sensor). The wavefront curvature can be reconstructed by analyzing the local tilts. In advanced Shack-Hartmann architectures, the microlenses are replaced by ▶ microaxicons to improve the robustness of the system. Extended functionality is obtained by combing the Shack-Hartmann principle with ▶ autocorrelation setups (see also: ▶ wavefront autocorrelation) and spectral resolution.

SHG

▶ Second harmonic generation.

SLM

▶ Spatial light modulator.

Slowly varying envelope approximation

Approximation of sufficiently long pulse durations in ultrashort-pulse physics that enables to assume that interactions do not depend on the ▶ carrier envelope phase offset of the electric field oscillation.

SNOM

Scanning near-field optical microscope, also NSOM (near-field scanning optical microscope). High-resolution optical microscope based on sub-wavelength apertures (e.g. fiber-tips) or scattering tips (apertureless near field optical microscope). SNOM enable for detecting the fine-structure of optical fields, e.g. after propagating through micro- or nanooptical devices.

Solid Immersion lens

Numerical aperture and magnification of a lens can be enhanced by completely filling the space between lens and object by a high-index solid material. Thus, the resolution of microscopes and data storage systems can be improved beyond the diffraction limit of optical systems in air. With hemispherical solid immersion lenses, a theoretical enhancement factor of n (where n is the refractive index of the lens material) for the NA can be obtained. With superhemispherical (Weierstrass) systems with truncated spheres of a height of $r + r/n$ (where r is the radius of curvature), the maximum theoretical enhancement factor is n^2. [1999_22, 2001_29]

Space-bandwidth product

The space-bandwidth-product is a measure for the upper limit of information transfer by light propagating through a given optical system. It can be described by the number of pixels required for maximum information capacity and depends on the coherence properties (quantum statistics) of the light field; see also: ▶ time-bandwidth product.

Space-time-coupling

For a proper description of the propagation of ultrashort pulses at extremely short pulse durations, spatial and temporal variables can not further be separated because of travelling time effects. This fact is called space-time-coupling [2002_48, 2002_49, 2000_9].

Spatial chirp

Spatial change of the spectrum of a wavepacket in contrast to the ▶ temporal chirp; see also: ▶ spatio-temporal chirp.

Spatial frequency

A spatial frequency is the counterpart to the temporal frequency of a signal in spatial domain instead of temporal domain; see also: ▶ time-to-space conversion.

Spatial light modulator (SLM)

Pixellated optical device for a spatially resolved phase and/or amplitude modulation of an optical field. In many cases, liquid crystal displays (LCD) or digital micromirrors (DMM) are used, e.g. for data projectors and cinema projectors (▶ see also: digital light processors). SLM are also used for ▶ temporal pulse shaping.

Spatio-temporal chirp

Simultaneous change of the spectrum of a wavepacket (e.g. of an ultrashort laser pulse) in space *and* time; see also: ▶ temporal chirp, ▶ spatial chirp and ▶ space-time coupling.

Spatio-temporal coupling

See: ▶ space-time coupling.

Spectral interference

Interference of coherent spectral components of polychromatic beams, e.g. of ultrashort-pulse lasers. Spectral interference results in characteristic spatio-temporal interference patterns like ▶ X-waves or ▶ X-pulses.

Spectral phase

The result of the Fourier transform of the time dependent field of an ultrashort pulse can be separated into a part describing the spectral distribution of intensity (spectrum) and a term describing the spectral phase (analogous to the temporal phase) [2000_9, pp. 11-27].

Spectral phase interferometry for direct electric field reconstruction (SPIDER)

Method for the characterization of the spectral phase of ultrashort pulses by splitting the beam in two replicas, inducing a known spectral dispersion in one arm of the system and generating a detectable spectral beat signal by superimposing both.

Spectral transfer function

The spectral transfer function (STF) describes (in analogy to ▶ modulation transfer function or ▶ phase transfer function) the change of the spectral distribution of a beam, in particular for broadband sources like ultrashort-pulse lasers.

Specular reflectance

Part of the light reflected from an object surface in the direction given by the law of reflection of geometrical optics. Specular and ▶ diffuse reflectance can not completely be separated because of their overlapping spatial characteristics.

SPIDER

▶ Spectral phase interferometry for direct electric field reconstruction.

SPIRIT

Acronym for "spectral interferometry resolved in time" which is a method for pulse characterization based on a transverse shearing of spatially dispersed spectral components of a pulse [2003_28].

Strehl ratio

The Strehl ratio (SR) as the normalized peak intensity of the ▶ point spread function is a measure of the maximum possible light concentration in imaging optical systems.

Sum frequency generation

Nonlinear optical process resulting in a frequency conversion. The interaction of light with a material enables to combine the energy of two or multiple photons and thus to generate photons of higher energy. The resulting frequency corresponds to the sum of the fundamental frequencies. A special case of SFG is the generation of ▶ higher harmonics of a fundamental beam, e.g. ▶ second harmonic and ▶ third harmonic generation.

Superluminal

Nondiffracting beams can show effective superluminal behavior with respect to their group and phase velocities [1997_42, 1999_13, 1996_7, 2001_18]. This was experimentally verified by ionization experiments [2002_27].

Surface spatial wavelength

Wavelength of surface structures in spatial domain, see also: ▶ autocovariance and ▶ correlation length ([1993_6], pp. 44-47).

Talbot effect

▶ Self-imaging of periodic phase and/or amplitude objects by constructive interference [1836_1]. In certain distances (Talbot distances), the original phase and amplitude pattern is replicated, also in fractional multiples of the Talbot distance (fractional Talbot effect). A more detailed analysis shows, however, that the replication of the initial field pattern can not be perfect. See also: ▶ Montgomery effect.

TBP

▶ Time-bandwidth product.

Temporal chirp

Change of the spectrum of a ▶ wavepacket (e.g. of an ultrashort-pulse laser) on the time axis during one and the same pulse; see also: ▶ spatial chirp and ▶ spatio-temporal chirp.

Temporal pulse shaper

Optical system for the temporal shaping of ultrashort pulses, in particular by spatially splitting a beam into its spectral components and multichannel processing these components with a ▶ spatial light modulator.

THG

▶ Third harmonic generation.

Thin-film microoptics

Microoptical components based on thin-film microstructures which can be fabricated by deposition techniques (e.g. vacuum deposition with shadowing masks, ▶ LPVD) or by poststructuring of deposited layers (e.g. lithographically); see also: ▶ thin-film microlenses, ▶ GRMMA.

Thin-film microlenses

Microlenses consisting of structured thin optical layers; see also: ▶ thin-film microoptics.

Third harmonic generation

Tripling the light frequency by nonlinear optical processes. Typically, ▶ second harmonic generation is combined with a subsequent ▶ sum frequency generation.

Time-bandwidth product

Product of temporal and spectral width of a pulse. In the special case of a Fourier-limited (bandwidth-limited) Gaussian-shaped pulse, the minimum time-bandwidth product is 0.44 for $sech^2$-Pulse and 0.32 for Lorentzian pulses. In general, it depends on the shape, symmetry and frequency ▶ chirp. See also: ▶ wavepacket and ▶ space-bandwidth product.

Time-nonstationary filter

A time-nonstationary filter can be realized either by an extremely fast sampling or a nonlinear process and is a necessary condition for the measurement of pulse durations in ultrashort-pulse domain [2002_29] .

Time-to-space conversion

Conversion of a temporal structure into a spatial structure [2002_35]. For example, the information about the pulse duration of a bandwidth-limited pulse can directly be measured by analyzing the spatial distribution of interference fringes generated by crossed beams in a single-shot ▶ autocorrelator setup.

Toraldo concept

Concerning to an idea of Toraldo di Francia, optical systems can be adapted to a given imaging problem by properly adapting the pupil function. Thus, the radius of the central disc of a diffraction pattern can be reduced and the surrounding dark ring can be broadened. The principle is currently applied to the separation of stars and extrasolar planets in astronomy.

Transform limited pulse

Pulse duration and spectrum of a transform limited pulse are directly related to each other by a Fourier transform because its phase function is constant.

Trap, optical

An optical trap uses the optical forces to fix and manipulate particles. With arrays of beams, multiple particles can be moved or rotated simultaneously.

Triple autocorrelation

▶ Interferometric autocorrelation with three instead two interferometer arms performing two autocorrelation measurements instead of one. The triple autocorrelation enables for revealing signal asymmetries even with a linear autocorrelation method but is more complicated compared to the conventional autocorrelation.

Truncated Bessel beam

Single maximum beam resulting from the truncation of a Bessel beam with a diaphragm. If the aperture diameter matches the first minimum of the Bessel profile, minimum diffraction is obtained (self-apodization setup, see also: ▶ apodization).

Undistorted progressive waves

Term introduced as early as in 1941 by Julius Adams Stratton in the framework of general electromagnetic wave theory [1941_1] for particular propagation-independent solutions of the wave equation. See also: ▶ nondiffracting beam, ▶ diffraction-free beams.

VRM

Variable reflectance mirror, see: ▶ graded reflectance mirror.

VUV

Vacuum ultraviolet. In contrast to the visible range, VUV light is absorbed in air. The short-wavelength edge of VUV is not clearly defined (between 5 and 190 nm). In the gap between VUV and the X-ray range, the ▶ EUV range is located.

Wavefront

The wavefront of a monochromatic optical field (or phase front) comprises the entirety of points of equal phase for which the corresponding ▶ Poynting vectors are normally oriented to the local tangential plane. For a polychromatic wave, subwavefronts can be defined for arbitrary wavelengths within the given spectral profile. The spectral phase of an ultrashort-pulse laser consists of a dynamic continuum of wavefronts. Propagation effects of short pulses can lead to pulse front effects which can not be described by the picture of a wavefront (see: ▶ pulse front). In general case (broad bandwidth, ▶ temporal chirp and/or ▶ spatial chirp), the wavefront is a complicated function of spectrum, time and/or space. See also: ▶ Hartmann sensor, ▶ Shack-Hartmann sensor, and ▶ wavefront sensor.

Wavefront sensor

Sensor for the detection of the spatially resolved tilt of a wavefront with interferometric, Hartmann, Shack-Hartmann, or axicon based Shack-Hartmann setups (see; ▶ Hartmann sensor, ▶ Shack-Hartmann sensor, ▶ axicon). The global wavefront curvature can be reconstructed from local tilt data with adapted algorithms.

Wavefront autocorrelation

Combination of ▶ wavefront sensing with ▶ collinear or ▶ noncollinear autocorrelator. Thus, a sequential or simultaneous measurement of spatially resolved ▶ autocorrelation and wavefront of an ultrashort ▶ wavepacket is enabled.

Waveguide

Microstructure which guides light by reflection at inner walls (hollow waveguide), by total internal reflection (refractive index step) or refractive index gradients (see: ▶ GRIN lenses).

Wavelength division multiplexing
Multiplexing of a signal by a spatial separation of different spectral components.

Wavepacket
Spatio-temporally localized (pulsed) solution of the wave equation corresponding to a certain spectral distribution. For a Fourier-limited (bandwidth-limited) pulse, the relationship between spectral and temporal width is directly given by a Fourier transform. For pulses of temporally variable frequency (see: ▶ chirp), the Fourier transform can not reveal the complete information on the complementary parameter.

WDM
▶ Wavelength division multiplexing.

Wiener-Kinchin theorem
The theorem says that the Fourier transform of the field autocorrelation is the spectrum of the time-dependent field (square of the magnitude of its Fourier transform). Therefore, the autocorrelation (without additional information) is not sensitive to the spectral phase.

X-pulse
Pulsed ▶ X-wave.

X-wave
Wavepacket with the shape of an "X" in the projection onto a two-dimensional plane in space-time coordinates. X-waves result from the ▶ spectral interference of ultrabroadband conical waves. Originally this phenomenon was described for acoustical waves [1992_2, 1992_3].

XPM
Cross-phase modulation.

Zernike Polynomials
Polynomials for the description of ▶ wavefronts and wavefront ▶ aberrations [1995_27, 1998_2]. See also: ▶ Seidel coefficients.

INDEX

A

Printed and bound by CPI Group (UK) Ltd, Croydon, CR0 4YY

08/05/2025

01864932-0001